The Complete Guide to Customer Support

Published by CMP Books
An imprint of CMP Media LLC.
12 West 21 Street
New York, NY 10010

ISBN Number 1-57820-097-0

April 2002

First Edition

For individual orders, and for information on special discounts for quantity orders, please contact:

CMP Books
6600 Silacci Way
Gilroy, CA 95020
Tel: 1-800-500-6875 or 408-848-3854
Fax: 408-848-5784
Web: www.cmpbooks.com
Email: cmp@rushorder.com

Distributed to the book trade in the U.S. and Canada by

Publishers Group West
1700 Fourth St., Berkeley, CA 94710

Design by: Saúl Roldán

Manufactured in the United States of America

Dedication

To Freddie Golino, the IT Manager at CMP's 12 West 21st St. New York office, where *Call Center* and it's sister publication *Communications Convergence* is published, and where CMP Books also has offices. Without *his* fine, friendly and patient tech support, on premises and to remote locations like Brendan's home office in Canada, this book would not have been possible.

Contents

Foreword:
Lending Support

BY WARREN HERSCH

The book you're reading is the direct descendant of <u>The Complete Help Desk Guide</u> a book by Mary Lenz, who preceded me as editor-in-chief of *Call Center* Magazine.

That work, published in 1996, is still around. And, as with this book, you can buy it from CMP Books (www.cmpbooks.com). I highly recommend it as a companion to this volume.

So why have Joe Fleischer and Brendan Read written a new title on support?

A lot has changed in six years. Support is no longer limited to the work of employees in corporate departments who fix problems with colleagues' computers. Instead, support encompasses the long-term care that organizations provide customers after they buy products or services.

Any company can provide support; it's not just for high-tech anymore. You can expect support reps at businesses that sell or repair PCs and printers. You might not expect support reps at businesses that offer travelers' insurance.

But at either type of company, customers are getting in touch with people who can answer questions about products or services that customers have paid for.

These companies are also offering automated hotlines and Web sites for customers who don't have time to speak with live reps, or who have the same questions other customers have already asked.

Customer support is an exciting field to be in. Compared to the pay in most call centers, the wages for people who work in support are significantly greater. As a result, so are the costs.

But the opportunities are well worth the costs. Some companies charge for support, thereby generating revenue. But even those that don't charge for support are enabling companies to establish reputations and retain customers. In good or bad times, keeping customers is what keeps you in business. Given two similarly-priced items of comparable quality, customers select — and stay with — the one that comes with superior support.

This book is a handy guide in many areas of support. Turn to any chapter on a topic, and you'll find helpful background combined with good ideas. If the writers occasionally restate certain points or definitions, all to the good: You remember them.

There's a lot to remember. Way back when, you were an expert if you knew what a trouble ticket was. Now support has an influence that extends beyond the help desk.

Consider: Customer relationship management (CRM) and knowledge management, respectively, are two practices that are changing how companies gather information about customers and present information about themselves.

People have written entire books on these subjects, including Janice Reynolds, the editor of this book. And where did CRM and knowledge management originate? You guessed it.

The methods that help desks used to document requests for support are evolving into collective efforts throughout entire companies to share what they know about customers. The problem resolution tools that help desks used in the early and mid-1990s, coupled with the ubiquitous search engines that ushered in the Internet age in the late 1990s, are maturing at the start of this decade into full-fledged knowledge management strategies.

Customer support operations are still call centers, after all, and you can expect to encounter many of the challenges of managing them. Among these: finding the right people; identifying the best locations; deciding which outsourcers can help; and looking for staff to assist from overseas.

You're in good hands. If there is a dynamic duo in the call center, it's Brendan Read and Joe Fleischer. As the editor-in-chief of *Call Center* magazine (the most-read and the most respected publication in the business), I can tell you that these writers complement one another very well.

Brendan, the author of <u>Designing the Best Call Center for Your Business</u> is a renowned authority on all facets of call center services. He regularly covers outsourcing, site selection, staffing and training in his longtime role as the magazine's services editor.

Working from his home in Victoria, BC, Canada, he has firsthand insight into teleworking. But he can also write about technology, which he did when he began with us. He still turns out an occasional story on technical subjects like multi-site call centers.

As you can see in this book, Brendan relates technology to call center services. And, as you'll discover from experiencing his inimitable style, he has a great sense of humor.

Joe, the magazine's chief technical editor, writes about many of the essential technologies in call centers, like routing, reporting, scheduling and call monitoring. He also discusses key practices such as CRM, knowledge management and quality assurance. He's also the first to admit he's not quite as funny as Brendan.

Together with Brendan, Joe has interviewed hundreds of managers of support operations. Both frequently contribute their knowledge as speakers, moderators and in-house experts at numerous conferences, including our own Call Center Demo and Conference. Brendan, for example, helped develop and serves as a panelist at CMP Media's annual Site Selection Summit at our spring call center show in Orlando, FL. Joe has created conference sessions for the Orlando event and for other call center shows, which you can find out about by visiting www.cmpevents.com.

Enjoy this book. Look at the case studies. Learn from support centers in leading companies like Linksys and Xerox. Discover the advantages of employing support reps who work from their homes. Read how tiers of support can help you. Find out how to recruit, hire and train the people who communicate with customers. Get the skinny on the strengths and weaknesses of live support versus self-service.

It's all here. And I'm pleased to lend my support to the culmination of Brendan's and Joe's long hours of researching and writing <u>The Complete Guide to Customer Support</u>.

— **Warren Hersch**
Editor-in-Chief
Call Center Magazine
April 2002

Preface

Support is the ongoing assistance a company provides customers with regard to its products and services. It's one of the most important — and challenging — job in any company.

Many of the books you're likely to find on support focus either on technology or management philosophies for help desks. This book is different in two ways.

First, our emphasis is on operations that support a company's customers outside a company rather than help desks that support colleagues within a company.

We recognize that help desks and customer support operations do have many similarities, especially when they provide technical assistance. In our suggestions for further reading, we refer you to some excellent books on help desks. These books cover areas of responsibility unique to internal support, like how a help desk can catalog and track changes to its company's equipment and software.

Compared to help desks, customer support operations bear a greater responsibility: they enable a company to earn the trust that's necessary for repeat business and provide the company with another revenue stream. In this book, we cover issues you're likely to encounter when you assist external customers.

Second, we broaden the definition of support to include a variety of products and services from cars, computers and appliances to insurance and financial services to health care organizations, home security and more — all of which entitle customers to a dedicated team of people who are available to answer their questions. Hence the title of this book: The Complete Guide to Customer Support.

Our goal is to assist you with developing a coherent strategy for supporting customers based on a combination of essential tools and practices. We describe the tools you need and we outline practices for applying them based on real-life examples.

Even more than in sales and customer service, support is where individual skills come into play. The people who work in support have to be experts in your company's products or services, whether they deal with customers directly or contribute to an on-line knowledge base that customers search themselves. This book shows you how to go about recruiting and hiring support reps, including the pros and cons of outsourcing support. We also describe how support centers can retain reps by offering incentives and options like working from home.

Numerous assessment and certification programs for support reps, supervisors and the centers themselves have emerged within the past decade. Ostensibly, the objective of the certification program for the individual is to make it easier for a company to identify applicants who have the skills you're looking for. We give you straight talk on how much impact certification should have in evaluating the qualifications of candidates for jobs in your support center and whether achieving certification for your center is worth the effort.

Most support occurs over the phone, and an increasing amount is happening on-line. The Internet makes it possible to support customers in ways that were unheard of ten or 20 years ago. In addition to being able to reach support reps via e-mail or text chat, customers who experience problems with their computers can use software to diagnose and fix them. Support reps can use software to take control of customers' computers, either to take care of technical problems immediately or demonstrate to customers how to use their computers, fill out a Web-based form or navigate the company's Web site. This book covers these and other technologies that are essential to support.

As we explain throughout the book, you can automate some aspects of support to reduce the amount of time customers spend waiting to speak with live people. For example, you can gather together information about your company's products and services within a knowledge base that reflects customers' most frequent questions, and you can enable visitors to your Web site to look up information from your knowledge base. You can also give customers the option of calling automated systems where they hear answers to their questions, all culled from your knowledge base, as synthesized speech.

However you choose to automate support, you need a team of knowledgeable people to ensure the information within your knowledge base is current and accurate. Our book furnishes you with the background to help you work toward an optimal mix of live and automated support.

Good support depends on the practices that underlie your use of technology. To run an effective support operation, you have to make the most of the reps' skills to apply the knowledge you collect about your customers. That is why we highlight practices as well as technology. These practices include customer relationship management, where companies learn more about customers so that they can serve them better each time they communicate with them. We also present examples of knowledge management practices that make it easy for customers to learn about your company and its products and services, yet allow support reps to retrieve information concerning questions to which they don't already know the answers.

To illustrate what it takes to develop reps' skills, we profile quality assurance efforts used within support organizations. We cover technologies like call monitoring and practices like assessment, evaluation and training.

Companies that stay in business don't just seek customers for the short term. They are aware that it's far easier to get customers in the door than to keep them coming back. They recognize that they need more than fancy technology or smart, hard-working employees; they have to provide long-term care that ensures the loyalty of their customers. That's the essence of support.

We hope you find this book helpful in mapping out an approach to support that welcomes your customers and encourages them to come back often. We wish you well in your efforts, and appreciate your comments and questions.

Joe Fleischer
Brendan Read

Acknowledgments

This book would not have been possible without Mary Lenz, the former editor-in-chief of *Call Center* magazine who set the stage in 1996 when she wrote <u>The Complete Help Desk Guide</u>. As the first editor we worked with on the magazine, Mary contributed greatly to our broad understanding of the intricacies of call centers. We also appreciate the guidance and encouragement of the magazine's current editor, Warren Hersch.

We extend special thanks to all the people whom we had the opportunity to interview in preparing this book. They represent a cross-section of the customer support community — consultants, support managers, leaders of support organizations and support vendors. We are pleased to share their knowledge and experience with you.

There are several individuals who have helped us gain greater understanding of support. They include Al Hahn, Hahn Consulting; John Hamilton, Service Strategies Corporation; Rick Kilton, RWK Enterprises; Mia Melanson, Performance Consulting; Bill Rose, founder and ceo, the Service and Support Professionals Association; and Fred Van Bennekom, Great Brook Consulting.

In addition, the authors would like to thank the team that produced this book. Janice Reynolds, who edited Brendan's previous title, <u>Designing the Best Call Center for Your Business</u>, deserves kudos for coaxing us through this book's completion. We were fortunate to have the design expertise of Saul Roldan, the longtime art director for a number of CMP publications. The business acumen of Matt Kelsey, Christine Kern and Frank Brogan has enabled us to bring this book into your hands.

Finally, we can never offer enough appreciation to our spouses and families, whose kindness, wisdom and strength make everything worthwhile.

Introduction

Ever walk into a store and find that almost no one knows how to help you? Particularly in stores that sell high-tech products, the people who are best able to answer your questions aren't necessarily where you make your purchases.

Consider: Linksys, which sells computer networking equipment, is adding support staff to serve a growing number of home and small-business customers who depend on the vendor, rather than the mega-stores that stock the products, to be the primary source of support.

The demand for support is substantial. Linksys' products occupy at least 30% of the shelf space devoted to networking equipment in retail stores, says CEO Victor Tsao, citing statistics from market research firm NPD Intelect. The Irvine, CA-based company, which has offices throughout Europe and Asia, receives a million support calls a month from US customers alone. Since Linksys doesn't sell its products directly, the calls come from customers who buy the company's products from retail stores, retail Web sites, resellers and high-tech catalog firms, including CDW and MicroWarehouse.

Yet when the authors of this book spoke with Linksys at the end of February 2002, the company's in-house support operation at its headquarters comprised between 80 and 90 support representatives. And these representatives, or "reps" as we call them throughout the book, answered calls only on weekdays.

So how is Linksys able to assist its customers after they purchase its products? That's where this book can help you.

Whether or not your company sells high-tech products, The Complete Guide to Customer Support covers all you need to know to answer this same question for your support center.

This book is the result of hundreds of interviews with managers of support operations, as well as with consultants and vendors. The goal of our research is to help you see the big picture by recognizing the technique behind the brush-strokes.

To start off, we provide the background to determine if you need to set up or expand your company's current support operation.

Linksys is fortunate in that its CEO, Mr. Tsao, knows the value of support firsthand. At his previous job with Taco Bell, he was responsible for providing technical support to employees on equipment and software.

Linksys also has clear reasons for growing its support center: the company's products have become mainstream. Karen Sohl, Linksys' manager of communications, points out that in 1998, when Linksys had as few as 60 employees, the company usually dealt with technically savvy customers who knew something about data switches, data hubs and network interface cards.

But as phone and cable companies continue to roll out high-speed Internet services for consumers and small businesses, Linksys has begun to tap more into what Sohl calls the "mom-and-pop market."

Once you establish demand for your products, as Linksys has, the next step is to hire people to support them. This book guides you through finding the right staff for your support center.

Given its widening base of customers, Linksys plans to hire at least 100 more people, mainly in support and customer service. Linksys mainly recruits from trade schools rather than seeking college graduates with computer science degrees. New support reps, who begin as hourly employees, earn the equivalent of between $20,000 and $30,000 per year.

As you would expect, reps with more experience and skills earn more money, according to the hierarchy in support centers. This book outlines what skills reps need to climb up the hierarchy, and what criteria are best for assessing applicants for positions at different levels in your support center.

As this book also explains, you don't necessarily have to limit your choice of staff to the local workforce or even within your own company. For example, Linksys employs 45 support reps from an office in Manila, The Philippines, to assist customers by e-mail. You, too, may have considered opening a center overseas; we present the pros and cons of doing so. We suggest other options, like employing teleworkers, if you want to add staff but don't have the opportunity to move to a larger site.

Another way to bring in more staff is to hire one or more firms to answer customers' questions on your company's behalf. We explain how these firms, which we call "service bureaus" or "outsourcers," can manage some or all of your company's support operation.

Several outsourcers help Linksys' customers after hours (between 5 pm and 6 am) on weekdays and all day on weekends. The outsourcer with the largest assignment is Microdyne, which has a support center near Los Angeles International Airport in Torrance, CA, where some 330 support reps answer calls from Linksys' customers.

The site is within driving distance from Irvine, which means that support reps from Microdyne can easily acquaint themselves with new products at Linksys' labs.

For reps within and outside Linksys, such knowledge is invaluable because products continually change. That was certainly true in 2001, when Linksys revised 50 of its products and introduced new wireless networking devices.

But support reps don't always have to answer inquiries from customers. Linksys has automated systems in place that let customers find answers to some of their most frequent questions by themselves.

One of these automated systems is a searchable knowledge base on its Web site, which Linksys maintains using iKnow software from Peregrine Systems. Visitors to Linksys' Web site type in words to describe the products or categories of products they have questions about. The Web site displays a list of possible answers, and after visitors select answers, they can indicate if the information has helped them. The Web site then displays a page that allows customers to send e-mail messages to Linksys if they have additional questions.

The book tells you about on-line knowledge bases and other tools for automating support, including interactive voice response systems that customers can call to hear information about your company. Plus, we devote an entire chapter to the issues you're likely to face when establishing a knowledge base, such as who contributes to it.

Knowledge management is one area where technology and practice intersect. Another is customer relationship management (CRM), which refers to information companies consolidate about their customers.

From a technical standpoint, CRM begins at the simplest level with a method of tracking support requests. Linksys, for instance, uses iHeat software from FrontRange Solutions to allow Web site visitors to submit on-line forms with inquiries about specific products.

But the practice of CRM is more complex than installing software. We describe how CRM enables your support center and the rest of your company to become collectively more knowledgeable by sharing information you've all gathered about customers.

We also highlight other technologies that are essential to your support center, including call routing, reporting and monitoring.

Technology is a great instrument but it means little if you don't have the staff to use it. Example: Victor Tsao, Linksys' CEO, says the company spent $1.6 million on a new 12,000-port phone switch from telecom manufacturer and developer Avaya. Yet, as one small business customer of Linksys has told us, callers to Linksys' support center occasionally have to leave voice mail messages if live support reps aren't available. Voice mail is not acceptable in any call center, let alone in a support center that has no way of discerning in advance which calls are urgent. The phone switch's call routing capabilities are only as good as the people who pick up the phones.

But Linksys is heading in the right direction, and Mr. Tsao is demonstrating his commitment to support by ramping up Linksys' staff and technology. To accommodate the additional in-house staff the company intends to hire, Linksys shifted its customer support and customer care operations from its headquarters, which had 300 employees prior to the move at the end of December 2001, to a larger facility in the same city.

The new site, a former clothing warehouse two miles away from headquarters, should provide Linksys with plenty of space for new support staff, as well as for product labs and the new phone switch from Avaya.

There's a good chance the support reps who are currently with Linksys will stay around to train the new recruits. Annual turnover among support staff at Linksys averages between 10% and 15%, and the average tenure among reps is two to three years.

This book navigates you through the challenges of building an environment that encourages support reps to remain with your company. Among the issues we cover are choosing sites where employees can easily get to work, and designing comfortable, ergonomic facilities.

We also stress the importance of offering support reps career paths. At Linksys, for example, more than half of the members of the product management team began their careers with Linksys in support.

Lastly, we point to key trends in support, such as benchmarking and certification, which aim to give support center managers some standard to judge the quality of their efforts. We also look at opportunities to generate revenue by charging for support, which many companies, Linksys among them, have opted against. In evaluating this tactic, we describe the circumstances where it does make sense for your business.

As with all books CMP Books publishes, we offer a list of sources of additional information, including other books we recommend on support. Beyond this book, you have a wide range of places to go to further your education in support, such as *Call Center* magazine, where both authors serve as editors, and regular events like Call Center Demo and Conference.

One

CHAPTER ONE

WHAT ARE CUSTOMER SUPPORT CENTERS?

What Are Customer Support Centers?

C H A P T E R One

Every customer needs access to service, but not every customer needs access to customer support. Similarly, while every product or service needs customer service to ensure a satisfied customer base, not every product or service needs customer support.

To decide if you need a customer support center, you must determine if the product or service that you provide requires it. And to do that, you have to examine the difference between service and support.

Service is a universal term that encompasses most calls from customers. Service can include requests for general information about a company, like hours of operation, store locations, mailing addresses or Web site addresses. These types of calls can come from customers even if they do not purchase anything from your company.

If customers do buy from your company, service then involves a whole other set of calls. Customers who have questions about processes related to purchases, like dates that packages are supposed to arrive or logistics for returning items, often speak with agents in customer service. Callers also reach out to customer service to praise or complain about how they feel the company treated them.

But when customers have questions, suggestions or problems about the products or services your company provides, you enter the realm of support, which is another galaxy in the universe of customer service.

Support centers, unlike other kinds of call centers, focus on solving problems, not just on providing general customer service. There are a variety of ways a company can provide support. They can staff the support center themselves, or they can enlist the help of outsourcers.

While it's vital that you man your support center with qualified support reps, many companies find that it's also vital that their support centers give customers automated options for finding solutions to problems. Examples of these include interactive voice response (IVR) systems, Web sites and automated replies to customers' e-mail messages.

When a customer brings a problem to the attention of a support center, a supervisor at the center opens a case and assigns a support rep to the case. The rep's job is to resolve the problem, preferably to the customer's satisfaction, before closing the case.

Support centers use problem management software to open, update and close cases. When reps or supervisors open cases, they create a trouble ticket for each case. Trouble tickets contain information about the cause of the problem and about the customer who identified the problem.(More about technology in chapter 7.)

Each trouble ticket has a unique number that enables support reps and customers to track an issue. Some companies even allow customers to view their trouble tickets via a secure Web site. That creates an expectation among customers and colleagues that support reps will regularly provide updates to the trouble tickets concerning their ongoing efforts to resolve customers' problems.

Trouble tickets serve another purpose beyond documentation. They are tools to help ensure that support reps take ownership of problems until they solve them.

Discovering the Diagnosis

Like medical files, trouble tickets describe attempts to diagnose problems so they don't recur or so that they have minimal impact if they do recur. Sometimes the cure is a replacement for a product that's broken. Other times, the cure comes from a recommendation from a live rep or a document on a Web site on how a customer can fix a problem and prevent, or at least work around, future problems.

Support reps often have to decide if a customer's problem results from the company's product alone or from the product's interaction with other items, usually from other vendors, which the company does not support. A typical practice is for the support rep to ask the customer a series of questions to identify possible causes. The more variation among your company's products, and the greater the number of distinct factors that affect how well the products work, the more questions your reps will need to ask.

A challenge that many support centers encounter, whether customers' difficulties are with medications or computer systems, is to how to narrow down causes of problems quickly and accurately.

One approach is to use a system that automatically presents customers with questions to help everyone find out the most likely reasons for the problems, and if possible, explain to the customers how to resolve them by themselves without the intervention of a support rep.

Before the advent of the Internet, computer manufacturers like Dell and Gateway primarily directed customers to IVR systems for automated support. Today these systems are among several options provided to the customer seeking support.

For example, the corporate Web site has been added to the mix; thus enabling self-service via on-line knowledge bases, automated diagnostic software and downloadable software patches. But IVR systems continue to be a widely available way for customers to receive support without the necessity of waiting on hold for a support rep.

When customers call a company's toll-free numbers for automated technical support, they reach an IVR system that plays prompts that tell them to respond to questions by selecting buttons on their touchtone phones. The questions from the IVR system ask customers to describe their problems or questions. For instance, if a customer has a problem with a computer, the IVR system asks the customer whether the issue is with software, the computer itself, or a piece of equipment connected to the computer.

Also, if you charge for support (more about that in chapter 3) the IVR system can ask customers for their account number and connect them to the right rep. Some premium level support plans promise dedicated account reps and no queues.

> ✦ **NOTE:** This information no longer has to come only from the customer. Customers' computers now often include software that automatically sends technical details to the support center (see chapter 7).

After the IVR system collects sufficient information from a customer's answers, it suggests probable solutions. As with all IVR systems, callers hear either recordings of actual people or computer-generated voices, but the ultimate source of both the questions the system asks, and the solutions it offers, is a knowledge base. Customers are gaining more firsthand experience of knowledge bases now that they have the choice of looking up solutions to problems on Web sites in addition to calling for automated support.

We discuss automated options such as IVR systems in chapter 3 and on-line knowledge bases in chapter 10.

Another area of your operations that affects how quickly your center can find causes of customers' problems is your center's approach to routing calls to reps. The way your center directs calls, in turn, depends on the groups you assign reps to. Support centers usually organize reps into groups based on the amount of training, experience and expertise they have.

Even the best support reps will run into problems that can't be resolved over the phone. But they can eliminate all known possible causes why the !#$%^&&*() device isn't working before recommending that the product be shipped to a repair depot or before they call field service.

You cannot expect *any* rep, no matter how much he or she knows, to solve *every* problem. Nor can you expect a new hire, especially one with little experience, to help a customer with a complex problem. That's why 'levels' are a common hierarchy in support centers.

As in most call centers, first-level or level-one reps, those with the least experience or the least specialized training, tend to be the first to speak with customers. If a problem is highly technical or it's simply one that a first-level rep cannot solve, a customer speaks with a higher-level rep with more experience, training and/or expertise in specific areas. We describe the strengths and weaknesses of this hierarchy when discussing levels further in chapter 4.

Your center's method of routing calls says a lot to customers. Let's say your callers reach an IVR system that doesn't attempt to solve their problems, but instead asks callers which products they need help with before routing them to support reps. Already customers should have figured out two things:

If they want assistance with specific products, they have to reach a certain group of reps. More importantly, customers need to be sure the center they are calling actually supports the products they're calling about. We examine the implications of routing in chapter 2.

So far, we've explained automated support and routing separately, but both have to work in tandem to work at all.

One reason is that IVR systems generally allow callers to reach live support reps if the automated answers aren't enough. Another reason is that callers occasionally need assistance from more than one rep.

Let's say, for instance, that neither the first rep the customer speaks with nor an automated system is able to help a customer resolve a problem within a certain amount of time. In this situation, the rep or a supervisor escalates the case to the next support level until someone at the center achieves a resolution.

Keep in mind that a resolution means that the support center closed a customer's case, not necessarily that it fixed the customer's problem. There are problems that a support center cannot solve by itself, such as when a company sells defective products.

If you're thinking about your company as a whole, closed cases with unresolved problems can be blessings in disguise. The support center emerges as the vital front line in quality control; and also as an interdependent link in the chain of relationships between your company and your customers. If many of the same problems come into the support center, that's a cue to your company that it has to correct these flaws quickly.

Fixing Problems While Satisfying Customers

Those of us who have engineers and techno-geeks in our families know all too well that such individuals can seem like alien body snatchers. They look and feel human but they act as though they're from another world: a world of circuits, electronics and moving parts that function precisely according to the iron laws of physics.

But whether a company builds machines or offers financial services, knowledge alone is not the only qualification for a support rep. As reps pinpoint the causes of customers' difficulties or answer questions, they have to be careful how they communicate with the customers. Unless customers clearly violate the terms of their warranties, reps have to avoid even the appearance that they're judging customers' intelligence or competence. Reps work for customers, not the other way around.

For instance, when a problem lies with the customer, and not with the product, the support rep needs to demonstrate tact and a desire to guide the customer to the best resources, not just the ability to solve the problem.

Co-author Brendan Read knows this well from his experience as a customer. Like many customers he does not have the best computer skills but unlike many customers he admits it (he dropped a computer programming class in the days of batch-job-processing, punchcards and PL/1).

In the fall of 2001, he moved back to Victoria, British Columbia, Canada, from New York City. In New York, he had dial-up Internet access from Mindspring, which Earthlink subsequently acquired, and digital subscriber line (DSL) service from Verizon. Both worked beautifully.

After he moved, he could not get either his new dial-up or new DSL service to function well on his desktop computer, even though both services worked fine on his company-issued laptop. He also installed a free downloadable variant of ZoneAlarm as a firewall, but was forced to disable it every time he connected to his company's virtual private network.

He talked to many support center reps, and he changed his desktop's settings innumerable times on their advice. After hearing that firewall software like ZoneAlarm could cause problems, he tried to delete the software from his desktop. He also bought file remover software to weed out unused files. Nothing worked.

So our intrepid co-author contacted a support rep from Telus, the local carrier that provided the DSL service. The rep explained that the problem could lie in the firewall software's settings, which varied among carriers, and recommended getting in touch with a consultant.

Read followed the advice and brought in a consultant from CompuSmart, a local firm, who cleared out all the software on the hard drive and then reinstalled it. The consultant suggested that Read stay away from firewall software and buy firewall devices instead.

This story illustrates a unique characteristic of support centers. They require reps who possess an arguably contradictory mix of customer service and problem-solving skills. This set of skills has an impact on all areas of setting up a support center, such as how your company chooses sites, designs facilities, selects technology, recruits staff and trains reps.

The moral of this story is that the customer has the ultimate responsibility to listen to the advice from the support rep and take the right steps to make their product work. That's why reps require not only knowledge, but also communication skills to earn customers' trust in their suggestions for solving problems.

What Makes Support Different?

Customer support is different from customer service in that in addition to good communication skills, support reps need a combination of knowledge and problem-solving skills. There also is often a greater degree of trust involved in support than in service because customers are relying on the reps to resolve issues that they believe to be urgent, regardless of whether customers are fully informed about the products or services they're asking about. A support rep must be able to reassure a customer as well as diagnose a problem; the rep is both a doctor and a nurse.

Qualifications for a support center rep can vary. In some centers, applicants have to pass tests that demonstrate aptitude in solving problems and serving customers. Other centers prefer or require specific bachelor degrees. Support centers often seek reps who have earned certificates from certain vendors or from organizations like the Services and Support Professionals Association (SSPA) or the Help Desk Institute (HDI). There's more about this in Chapter 8 where we discuss the issues of staffing and training.

Frederick Van Bennekom, principal with Great Brook Consulting in Bolton, MA, teaches operations management at Northeastern University and has written extensively on customer satisfaction. He explains the differences this way:

"The biggest difference between customer service functions, like order taking, fulfillment and general information and customer support is that there is a much greater degree of uncertainty with customer support.

"While customer support, as well as customer service, rely on computer knowledge bases, there are many more and much more complicated questions posed in support whose answers are not in the knowledge bases. Customer support agents must be ready to respond to those unanticipated questions."

Do You Need a Support Center?

As with call centers in general, support centers defy categories. A support center can employ one or 1,000 reps in Delhi NY; Delhi, Ontario, Canada; or New Delhi, India. You can designate a particular group of employees as support reps. Or you can establish an informal operation where anyone who picks up the phone at your company provides support. Your company goes with whatever works best with the volume of support requests it receives.

You pay a premium for expertise whether you run a support center for your

Support centers are usually practical and often not fancy facilities. Compaq's Ottawa, Ontario, Canada support center, located in a converted Beaver Lumber store, provides technical support for consumers who bought PCs from Compaq. The center has won company awards for its customer support. Set up originally to serve Canadian customers the center's excellent service led Compaq to route US calls there.

company or you're in charge of an outsourcer that handles support for clients. The more specialized the support, the more costly it is.

Sources like The Gartner Group's (Stamford, CT) report, "Contact Center

Support is Not Just for High-Tech Firms

Ask most people what support entails, and they'll tell you it's only for high-tech products and services, like computers, Internet access, network equipment and phones. Not so. It can apply to almost any product or service, like roadside assistance or insurance.

Problem solving and support is as necessary for callers stranded on highways as it is for callers lost on the Internet. A support rep frequently has to make similar decisions in either circumstance, like whether to explain to a customer how to restart a machine, be it a car or a computer, or dispatch someone to relieve the customer of having to touch the machine at all.

And sometimes it is just as important as having a non-technical support rep, or a 'level-zero' rep (see chapter 8) to field support calls based on priority and how valuable the customers is to the company (more about that in Chapter 11 on CRM).

Co-author Brendan Read remembers when he was a de facto 'Level Zero rep' with John Garde and Company, a small firm in Toronto that sold and provided support for industrial sewing machines. He also handled accounts receivable, invoicing, collections, shipping and over the counter orders.

"I would answer the phone and almost always the customer would ask for the owner, John Garde," he says. "If Mr. Garde was in — I would transfer the call to him. He would first diagnose and fix the problem on the phone. If he could not, then either he or one of the mechanics would go to the customer's site, or the customer would be asked to come in with the machinery. Customers from outside the Toronto area would ship the equipment in.

"If Mr. Garde was not in or not available, I would ask what the problem was, take down the information and tell the customer I would leave a message for Mr. Garde."

Self-Service Costs," say that call center customer service and sales phone calls cost $5.50 on average. But, support consultants say that cost for support calls range from $20 to as much as $100, depending on the length of the call and complexity of the problem.

For reps who handle basic support calls, such as requests to reset passwords, count on paying between 10% and 15% more than the average hourly rate at other types of call centers, says King White, vice president with the Dallas-based Trammell Crow Call Center Site Selection Group. For example, if you locate a support center where the average wage for a call center rep is $12 an hour, expect to offer at least $15 per hour for a qualified support rep.

For reps with the skills to provide higher-level support, including those who can fix customers' machines by connecting to them remotely, you're looking at much higher wages: between $20 and $25 per hour, depending on your location.

You're also looking at higher costs for real estate and facilities since support centers need more space to accommodate equipment and manuals than other types of call centers. White says that unlike the average call center, which can squeeze eight to ten workstations per 1,000 square feet (sf) per workstation, support centers can fit no more than six to eight workstations per 1,000 sf.

The expenses are worth it if offering support is essential to retaining the company's current customer base, generating leads for future sales and attracting more customers. Excellent support is not a cost but an investment in your customers.

But whether you need a support center depends on your business, and your customers. The more hands-on your customers are with your products and services, the more they need customer support.

Let's say your company makes power supplies and you mostly sell them through distributors like computer manufacturers. You need a support center to answer calls from customers who buy your power supplies direct from your company, but not from the computer manufacturers' customers.

Yet, if a computer manufacturer receives complaints about its products, and its support centers determine — correctly or incorrectly — that the cause of the malfunctions are with your power supplies, count on that computer manufacturer's customers burning up the phone lines at your support center.

Other Support Center Functions

The role of support centers is to solve customers' problems with products and services. But depending on the company and its products or services, more may be expected from individual reps.

Small firms typically need their reps to provide support and service; they don't have the luxury of specialization. For these businesses, support and service are one and the same thing. Support center reps also may handle internal

Cross-training Caveats

Customer support reps are smart people. They are very knowledgeable about your products and services. In theory, that makes these reps great candidates to sell.

One of the mantras of customer relationship management is that every communication with a customer presents an opportunity to sell. Outsourcers especially prefer, and often depend on, reps who handle outbound sales calls one week and provide support during the next. The idea is that the outsourcers reduce costs because they don't have to hire different people for different types of communication.

Outsourcers often have the resources, including software, that enables them to collect accurate information to cross- and up-sell to customers while providing support.

In practice, support and selling are two different skills, as outsourcer Stream International (Canton, MA) has learned. The firm broadened its services on behalf of clients to include cross- and up-selling after making its name in support.

A customer win-back program for one client, an Internet service provider, taught Stream that not every support rep was able to convince an unhappy customer to stay. The outsourcer had developed hiring profiles and training programs for sales, customer service and support reps.

"Some reps were successful at winning back customers but some were not," explains Deborah Keeman, Stream International's vice president of services marketing. "We had to put a separate profile together for the win-back program."

Berta Banks and Launa Green of the human resources consulting firm Banks and Dean (Toronto, Ontario, Canada, and Mequon, WI) have seen the consequences if you make support reps do jobs they're not qualified for, such as poor sales performance, higher absenteeism and turnover.

"You have to screen each rep for the attributes and skills for that other position, just as if you were hiring off the street," says Banks.

support because this function requires personnel with the same problem-solving skills and training. In either case, the demand isn't there to justify a dedicated support team.

Support reps can also cross-sell and up-sell products and service. But be wary about asking them to do it.

Training experts report that it is difficult enough for customer service reps to sell over the phone. Selling is a job that requires different attitudes and skills than serving customers, let alone providing support. Only if your company's product or service already includes support, such as supplemental insurance or an extended warranty, should support reps receive training on how to sell.

Van Bennekom points out that asking reps to sell to customers who call for support diminishes the credibility of the company and the support rep's advice.

"The support rep is the most out-there person in the company," he explains. "They identify much more closely with the customer than any other employee. That's what makes their bond with the customer. To ask them to start selling, especially if the customer is paying for that call, severs the relationship with the customer, and with the support rep and the company."

Peter Gurney, manager director of Seattle, WA-based reseach firm Kinesis recommends that companies first show rep how to apply what they know about customers.

"Cross-selling gives these agents an opportunity to deepen the relationship with these customers, as long as it is offered in the context of assessing and meeting their needs," says Gurney. "There is a risk that agents will anger customers by making inappropriate recommendations. Therefore companies should, instead of mandating cross-selling and up-selling, train agents to probe customer needs and make recommendations based on the information gathered from customers."

Integrating Support With Other Services

Customer support may work best for your company when you integrate it with other functions, such as sales, billing and repairs.

Perhaps the best example of customer support that's integrated with other functions is Ma Bell, the mother not only of all phone companies but also support. Most phone companies have separate numbers for repair, customer service and sales. But no matter which number a customer dials, the customer can receive help in any of these areas.

As noted earlier in this chapter not everybody makes all the components to a product or service. The problem the customer needs assistance with may or may not lie with 'the other guy's stuff.' Phone companies must deal with this dilemma daily.

Deregulation of the telecom industry is a boom in that it encourages innovation and drives down costs. But the downside with deregulation is that it complicates support since today's telecom customers can have separate carriers for local and long-distance phone service.

That leads to finger pointing when something goes wrong. And if you buy your own gear, you're responsible for it, from the service to the handset. Of course, you can pay the carrier extra for a wire maintenance plan, which can cut down on the number of calls you must make to determine the cause and resolution of your problems.

Co-author Joe Fleischer's experience should serve as a warning to companies that engage in finger pointing. When the office in New York City where he works switched to a new phone system in December 1998, each employee received a new direct phone number. All the phone numbers beginning with 212 were apparently taken, so the new phone numbers at the office began with 917.

The cutover to the new phone system at the office went perfectly. But if Fleischer's wife wanted to call her husband at his direct line, which was often the only option once the receptionists left at 6 pm, she had no way to do so. Why? MCI, the Fleischers' long-distance carrier, had phone switches that did not recognize 917 as a valid area code.

Fleischer spent hours at work and at home on the phone with Bell Atlantic (now Verizon), the office's local carrier, and with MCI trying to fix the problem. A colleague at his company in the facilities department offered her assistance, but to no avail. Neither carrier was able to solve the problem. When Fleischer tried getting help from MCI, a customer service rep told him that 917 was an area code reserved only for wireless or beeper numbers.

Then in January 1999, Fleischer followed a suggestion from one of Bell Atlantic's account managers: He tried calling the direct line using a phone company other than MCI. Sure enough, Sprint's phone switch recognized 917 as a valid landline area code. Needless to say, Sprint became his long-distance carrier. (He has since switched carriers again, but that's another story.)

To date, MCI is still soliciting Fleischer, as all phone companies solicit their competitors' customers, to receive its long-distance service. Fleischer is only one customer. So, despite MCI's apparent lack of awareness as to why he switched carriers, MCI still has plenty of other customers to keep it in business.

But there is also another lesson you can learn: support is to customer retention what avalanches are to mountain climbing. When a customer gets in touch with your company with a complaint or problem, every action your company takes from that moment on achieves greater magnitude.

If your company's products or services interface with others companies' products and services, cross-train support reps on the other companies products to the extent that's feasible, or work with the other companies to determine which organization handles support for which products. Communicate to support reps which products they support directly and which products they should refer customers to other companies for help. If you do this, you save your customer a lot of grief. Customers don't care who's at fault as long as the problem is satisfactorily taken care of.

Yet, one false move on the part of anybody at your company, like sharing inaccurate information or making comments that rub the customer the wrong way, and you risk losing that customer forever. You also risk losing others with whom the customer has any connection.

But when you give support reps access to correct information about your own company and its products and services, as well as relevant but thorough information about the people your company serves, reps are better equipped to make a coordinated effort to hold on to customers.

Justifying Your Support Center

As we noted earlier, providing customer support costs more than other functions like service and sales. Consequently, support center managers are constantly searching for new technologies, operational methods and locations to cut costs and boost productivity.

The same rationale for having a dedicated individual, individuals or facility for customer support is the same as for call centers in general. What are your customers worth? Can you afford to lose them? And when determining the size and scope of your support center, what is the demand?

Unfortunately many companies still treat their support centers as cost centers regardless of whether they assist employees or customers. They perceive customer support as an expense to be minimized.

These same companies don't recognize the relationship between support and keeping customers loyal. Instead, they see customer retention as the job of the marketing department. And they don't see the opportunities where they can recoup costs and make money from their support operations.

Firms can get away for a long time with this nonsense with their internal support centers, which are often unfairly named the 'helpless desks.' Unless internal productivity measurements prove that employees are not as efficient as they should be, and unless top managers scream that they can't get help, it's difficult to get companies to spend the money needed to improve internal support.

The overall consensus is that like water coolers and styrofoam cups, internal support simply adds to corporate expenses. Yet, if the company's employees can't do their job because of inadequate internal support, the company and its customers suffer.

Internal support is like the British 'Tommies' or infantrymen described in a famous Rudyard Kipling's poem *Tommy* about how nobody wants to give them adequate food or quarters or wants them around or gives them respect. Except when all hell breaks loose.

Which is why we're dedicating this book to our internal support professional, Freddie Golino, who is there for us, no matter where we are.

It's a bit easier to wrangle money out of the executive board for customer support. Customers buy products and services, especially computers, computer peripherals, and Internet access based on their perceptions of a company's support.

Admittedly, good customer support won't rescue a lousy product or service (even though there are companies in various industries that give the impression that market share dominance excuses poor products and mediocre service). But, good support usually will help competitively-priced, quality products and services to succeed.

Support is costly, but customers value it. Therefore, some companies, especially software vendors, have realized they can cover some, if not all, of their support costs and maybe even generate revenue by charging for support. (More about that in chapter 3.) By offering support at a premium, a support center becomes a profit center.

Two

CHAPTER Two

The Role of a Support Center

The Role of a Support Center

CHAPTER TWO

Before a company can provide assistance to customers who purchase its products, it has to be able to communicate the purpose of its support operation. This purpose should be clear to the people who provide support, and to the company's customers, especially those who pay extra for it.

What Does a Support Center Do?

Let's start with an academic description of a support center. The one below is an excerpt from a summer 1999 group assignment for students of the business school at Pace University in New York City:

> Your firm wants to establish a new call center to serve the NAFTA region. The company is a large service organization that repairs or replaces electronic entertainment equipment. The call center operator makes repair or replacement decisions based on the information that is called in from a database. The repair decision is based on a set of computerized rules.
>
> The application of the computerized rules will determine the action taken by the operator. If the item is to be repaired either the product will be returned to a regional repair depot or a local field service will be dispatched to the customer's location. The nature of the repair activity will be determined by the general business rules and by the type of service contract held by the customer.

The remainder of the project description indicates the number of calls the support center handles per month and per year, as well as the average length of a call.

To fulfill the assignment, the business school students drafted a policy statement, designed a technical layout for the center, chose sites, set staffing requirements and completed a five-year financial analysis. They also established timelines for each stage of the project.

In contrast to the hypothetical center from the business school project, most of the support centers we describe in this book provide help directly over the phone, and to a lesser extent, on-line. Reps at these centers don't necessarily send technicians to customers' homes or businesses; instead, their immediate goal is to help customers when they first get in touch with the center.

But whether they dispatch technicians or not, support centers usually share the same function — to resolve customers' problems or answer customers' questions. How support centers perform this function may vary; but the function itself does not.

Providing Prompt Support

Let's bring this academic exercise to life by looking at an actual example of the type of support center defined in the project.

Ever heard of Z prompt? If you own a printer or fax machine from Samsung and you visit samsungonsite.com, then you may have received help from Z prompt without knowing it.

Based in Irvine, CA, Z prompt is a service bureau that provides technical support, including on-site maintenance, to customers of companies that include Samsung, as well as support to employees of companies that include Mazda.

Through its support services, says Paul Musco, a vice president with Z prompt, the company troubleshoots Samsung customers' problems with faxes, printers and computer monitors, among other equipment. Z prompt also answers calls from customers whose Samsung computer monitors are under warranty.

At the start of 2001, Z prompt began to offer Samsung customers the option of asking for help by submitting on-line trouble tickets. This choice lets customers communicate their need for help immediately rather than asking them to wait on hold for reps to become available.

Z prompt uses Heat software from Colorado Springs, CO-based FrontRange Solutions to view information about customers and to enable them to submit trouble tickets from Web sites of Samsung and other clients. Z prompt also uses the software to create an on-line knowledge base to help customers with technical questions about Samsung's fax machines and printers.

To answer customers' calls and track on-line trouble tickets, Z prompt employs seven support reps who are available from 5 am to 6 pm Pacific time on weekdays. On average, these reps respond to up to 1,800 requests for support per month by phone and on-line.

Musco explains that about a third of on-line trouble tickets require calls from Z prompt's support staff to customers. By speaking with customers, the reps determine whether Z prompt should schedule maintenance or send field service technicians to their homes or businesses. Two of the reps serve as dispatchers and sometimes call customers to arrange for technicians to look at or repair their equipment. For on-site repairs, the company is able to rely on more than 1,000 firms that employ field technicians throughout the US.

Samsung's customers account for as many as 500 support requests per month. Besides getting in touch with Samsung about fixing computer equipment, some customers have warranties that allow them to exchange older models of Samsung's computer monitors for newer models from the manufacturer that have larger, flatter screens.

Customers can create on-line tickets at the Web site (www. samsungonsite.com) to request repairs or if they want to participate in this exchange program. Samsung passes these tickets along to Z prompt, which relies on the Heat software to keep records of these customers, including their addresses. The records enable local field technicians to know where to drop off new monitors for participants in the exchange program and pick up the monitors the customers want to replace.

After customers complete on-line forms to either seek technical assistance or to exchange computer monitors, the Heat software automatically sends them e-mail messages, which acknowledge their requests and provide them with corresponding trouble ticket numbers to refer to when communicating with reps.

On-line trouble tickets enable customers to request support as soon as they need it. But that fulfills only one half of the function of a support operation. Companies also have to set up systems to ensure they respond to customers, if not immediately, then within the period of time customers' support contracts allow.

The Heat software automatically alerts the two dispatchers if a rep at Z prompt does not refer a case to them within an hour after they've received a trouble ticket. Dispatchers also receive alerts if Z prompt has not resolved a case within the time specified by the client's service level agreement.

Customers can view the status of support requests by viewing on-line trouble tickets. But they don't always have to wait until they receive automated e-mail messages from the Heat software to get the help they need. Sometimes the reps at Z prompt can respond more quickly by calling them.

"If it's something we can potentially fix over the phone, we would call them first," says Musco.

For a growing percentage of Z prompt's customers, submitting on-line requests for help is often preferable to remaining on hold. Since the company introduced on-line trouble tickets, between 30% and 35% of customers have used them. Musco's goal is to increase that percentage to 65%. He's not far from meeting this target — as 2002 rolled around, a full 50% of customers were using the on-line trouble tickets.

Although Z prompt automated the way customers can inform reps of technical problems, the company still allows customers to receive immediate live help, if that's what they prefer.

"The way our phone system is set up, you're going to talk to a person," says Musco.

What Every Support Center Must Have

What's left unsaid in the academic project, but which is readily apparent to customers who call Z prompt or other companies for support, is that they need to be able to reach a live person, if necessary. Many times it's best if a customer can speak with a live support rep right away. Yet, the greater priority should be to ensure customers get in touch with the reps who can best help them.

The business school students at Pace, which included the cousin of one of the writers of this book, recognized that they had to keep their focus on the purpose of the support center. Only by emphasizing the purpose of the assignment — providing customers with an easy way to arrange repairs to the big-ticket consumer electronics items they bought — were the students able to complete every stage of the project, and earn an A.

Call Routing

A necessity in any support center, and in call centers in general, is that any reps with similar training should be able to assist any customers who require their skills. To make this possible, you need to devise a method for routing calls.

That's where your phone switch comes in. At a minimum, your phone switch should be able to direct calls to the first available support rep. The phone switch can include an automated call distributor (ACD) that allows you to define rules for sending calls to reps.

Up until the end of the 1990s, phone switch manufacturers used to incorporate ACDs as part of their equipment, but it's now possible to purchase call routing software that gives you greater flexibility in how you route calls than using your phone switch alone.

Let's discuss how support centers generally route calls to reps. Many support centers assign reps to groups that answer questions about specific products, provide services to certain categories of customers or perform particular functions. Support centers either provide customers with phone numbers they can call to reach these groups directly, or they initially direct callers to an IVR system that presents callers with a list of digits that correspond to a certain group. When a caller presses a digit, the call routing software retrieves the digit and determines where the call should go.

The term "skills-based routing" refers to the procedures we've just described. Strictly speaking, these procedures don't factor in agents' skills at all; they simply automate how callers connect with groups of reps. For instance, in the above scenario involving the IVR system, callers typically reach the first available rep in the groups they select.

But there are call routing software products that do actually allow you to assign a set of skills to a rep. These systems let you indicate primary and secondary skills for reps, or they enable you to create a list of skills and give each rep a score for each one.

The goal with this method of routing, which is closer to the skills-based model that many centers aim to follow, is that reps receive the largest percentage of calls that correspond to the skills for which they have the highest scores.

If two reps in a group are available to answer a call, but one has a better score for the particular product that the caller has a question about — based on the choice the caller made while listening to the IVR systems' prompts — the rep with the higher score is more likely to get the call.

Besides reps' skills, call routing software of today also considers the estimate wait times for callers who wish to get in touch with reps with certain skills or in certain groups.

Siemens' ResumeRouting software, which debuted in the mid-1990s, was among the first products from a leading phone switch manufacturer to offer true skill-based routing. Since then, more phone switch manufacturers have introduced skills-based routing software for customers who use their equipment. Avaya Business Advocate from Basking Ridge, NJ-based Avaya is a more recent example.

Quite a few software companies, including Oakbrook Terrace, IL-based Apropos Technology (which we mention later in this chapter), provide call routing software that works with phone switches from multiple manufacturers.

How Routing Relates to Support

True skills-based routing, as described earlier, is especially valuable and necessary in support because companies often offer different fee-based support levels (see chapter 3) and product lines and separate rep groups and experts. You can't train everyone to know everything well.

Here's how this works: a customer who bought a Plutonium Level service package that includes field support in 24 hours or less will enter their customer number at the prompt and is routed to a Plutonium Level rep. Or your firm makes Mac and Linux versions of the same software. The IVR prompt asks you your OS and you are routed to specially trained Mac or Linux reps.

Or you had a line of 386-implanted computers and you have still have a number of customers that won't part with them; today's reps would look at a 386 chip like a paleontologist would stare at a chip of dinosaur fossil "wow this actually worked?" So you find and train a group of compugeezers to support these machines.

As with skills-based routing, if the call and contact volume is slack for specific groups but crazed in others, you can route the calls outside the normal routing groups to handle overflow. But be careful. All of the reps in the overflow groups need to be trained, and, where required, certified (see chapter 8) to solve the problems of multiple products. While it would be inadvisable to route level-two calls to idle level-one reps, you can easily route overflow calls from level-one to idle level-two reps. (See chapter 4 on support center hierarchy.) Be advised that your top guns might not like flying subsonic, but it helps the cause, and, after all, that is what you're paying them for.

A common function within many companies' call centers is assisting customers who want to close their accounts. This function dovetails with another practice, which is establishing segments of customers, and creating groups of reps to serve these segments, based on how much or how often customers buy from the company. This, in turn, brings into play another routing method that reflects attributes of customers, not just attributes of agents.

In the realm of support, this means that a company as a whole must be knowledgeable about their customers' active warranties, service contracts, purchasing and repair histories and all problems the customers have experienced. The company then attempts to use all of this knowledge to hold on to their customers.

The more valuable the company perceives certain customers to be, in terms of the long-term revenue or repeat business these customers represent, the harder the company tries to keep them. It explains why, for instance, customers who call credit card issuers to empty their accounts end up speaking with retention specialists who try to persuade them not to leave, often through promises of higher credit limits or lower fees. We explore the issue of customer value, and the broader issue of customer relationship management, in greater detail in chapter 11.

How do support center reps become knowledgeable about customers? That's where information about customers comes into play. In some centers, reps are able to view customers' records on their computer screens at around the same time the call routing system directs customers to them. We describe how this process works in the next example.

Like the business school students at Pace, Daryl Breneman knows what it takes to fulfill the job of a support center. He is among a group of technical support managers for Primavera Systems, which develops project management software.

Primavera's main center is located at the company's headquarters in Bala Cynwyd, PA. Primavera also has a smaller center in Concord, NH, where 14 reps are available to help callers with questions about or difficulties with Expedition, a line of software from Primavera that tracks changes to projects.

In fall 2001, the center in Bala Cynwyd achieved Support Center Practices certification from the Service and Support Professionals Association (San Diego, CA) for its overall operation. In the real world, it's the equivalent of an A in customer support. (See chapter 9 to find out more about certification.)

One of the criteria that the certification looks at is how support centers use technology to run more efficiently. At Primavera, both the routing system and the customer information system work together to allow customers quick access to support reps, and to give reps quick access to callers' records.

That wasn't always the case. Primavera's support centers in Bala Cynwyd, PA, and in Concord, NH, have phone switches from Avaya, a phone switch manufacturer headquartered in Basking Ridge, NJ. The centers also use software from Bellevue, WA-based Onyx to look up customers' records. But the phone switch and the Onyx system did not communicate with each other so easily, which is why Primavera chose call routing software from an additional vendor.

Since the late 1990s, Primavera has used routing software from Apropos Technology, headquartered in Oakbrook Terrace, IL. Here's how Primavera uses this software. Callers to the company's center in Bala Cynwyd first reach an automated voice response (IVR) system that Apropos includes with its call routing software. Callers enter serial numbers for their project management software from their touchtone phones.

Primavera's routing algorithm uses the serial numbers to identify callers and find out if their companies are entitled to support. If callers are not eligible for technical help, Primavera directs them to sales reps who advise them on how to renew their company's support contracts. Otherwise, a caller reaches one of between five and six agents who ask callers whether they have questions about or problems with Primavera's software.

The agents then consult records on their PCs to see if a call is about a new or current support request. If the call refers to a current issue, the agent updates the record. If it's a new support issue, a new record is created. Next, the agents use Apropos' software to select which group of support reps should handle the call. The agents transfer the calls and the records to the appropriate support groups. From this point, customers who have been on hold the longest speak with the first available support reps in the groups to which the agents transferred them.

Between 54 and 55 support reps at the center in Bala Cynwyd assist customers with most of Primavera's products. Primavera assigns reps to groups related to certain products, and further divides these groups between reps who troubleshoot problems and those who answer questions about how to use products.

The center in Bala Cynwyd receives between 4,200 and 5,000 calls per month. Breneman says that customers often ask for help with installing Primavera's software and with printing project diagrams. The number of calls peaks after the December holiday season, which is when Primavera's customers start to plan their construction projects.

With this routing scheme, customers usually spend no more than ten minutes on the phone with the first group of agents at the main support center before speaking with a knowledgeable rep who can answer most of their questions. The same is now true for customers who contact reps at the center in Concord, NH, although they have to call a separate number to reach that center.

The company used to maintain a tiered or 'level' approach to handling calls (more about that in chapter 4), but that changed in July 2000, when Primavera established the routing rules described above (at the same time it moved its main support center from a smaller site in Bala Cynwyd). With its old routing scheme, Primavera required the first support rep the customer spoke with to answer that customer's questions within 15 minutes. If they couldn't, they transferred the call to a more knowledgeable rep.

"We still have levels, but for call routing purposes, we don't typically escalate calls," explains Breneman.

Expanding the Definition of Support

Unlike the reps at the support center described in the business school project, neither the reps at Z prompt nor the reps at Primavera necessarily rely solely on computerized rules to figure out how to assist customers. Automated escalations are in effect at Z prompt, and routing rules are used at Primavera to direct customers, namely those who are entitled to support, to reps with training on their specific products.

A support operation ultimately depends on its reps, not on computerized rules. If reps lack the skills to identify solutions to problems over the phone or on-line, automating certain aspects of the reps' job makes little difference.

The examples of Z prompt and Primavera illustrate the functions of many support centers, but, as we explain in the next chapter, support is not only about helping customers with technical difficulties. In its broadest sense, support is essential to any company that continues to value its customers after they've made their initial purchases.

Three

CHAPTER THREE

SUPPORT CENTER
OPERATIONS

Support Center Operations

C H A P T E R Three

On the surface, customer support centers appear not to be much different from other types of call centers. They employ people who receive calls, answer e-mail messages and conduct text chat sessions with customers, many of whom may have already looked up information by themselves on the company's Web site or by interacting with the center's interactive voice response (IVR) system.

But support centers are different in a few key respects. The calls and contacts can get hellishly complex and sometimes takes hours if not days to resolve, the solutions sometimes entail shipping the piece of !#$%^&*(() to the depot and having field repair to drop by and fix the wretched thingamabob.

For those reasons support centers are almost always more costly to operate than customer service/sales centers. Labor costs alone are 10%+. While you can offer as low as $7-$8/hour for a customer service agent you will have to fork over $20 to $25/hour for a senior rep. The costs of resolving a support issue, like 'my cellphone died' can range into hundreds of dollars compared with the tens of dollars to answer a customer service/sales issue: "Where can I sign up and get one of your phones. My old one didn't work and their tech support stunk."

But support centers can also make money on the service they provide. Support contracts have become, for an increasing number of firms, profit streams almost as valuable as the profit streams for the products. Like razor blades rather than the razors have become for razor manufacturers.

IBM is a good example. It jumped on that bandwagon, marketing their services as or more intensively than their machines. Their boxes seem to exist like the razor blades — to sell the services.

How Support Centers Communicate

Before we look at what makes support centers unique, let's look at what they have in common with other call centers. To provide good, customer-retaining support, while controlling costs and making money, support centers must select and use the best mix of communications methods.

Live Support

Live assistance, unlike on-line or automated assistance, is effective for most types of support inquiries. The more difficult the inquiry, the better the chances are that a live person can answer it.

The reason: there has not been a computer designed as effectively or as perversely as the mush, ooze and crunch of matter known as the human brain. Only the human biochemical computer can draw a relationship between an apple and an orange, unless it programs a nonhuman computer to do so. Only it can figure out that the problem is hardware not software, or software not hardware, or that the customer has blown a circuit, which is why it does not work.

At the same time only the brain would conceive of such self-destructive and utterly mindless activities like chucking apples and oranges at passing cars and trucks for fun or letting them rot in a vat to create booze to kill off more brain cells. Or like not checking to see if there is power to the machine, before calling customer support and screaming at the rep. No computer, unless programmed by a human brain, would be *that* illogical.

For customers, assistance by phone is the most effective method of communication that support centers can offer. It is much faster, simpler and more natural for a customer to talk to a support rep than to correspond on-line.

That's not to say that on-line assistance doesn't have its advantages. E-mail and some, but not all, text chat software furnish customers and reps with written transcripts of their communication. Also, many software products that deliver e-mail and text chat messages from customers to the reps enable reps to look up possible answers from a knowledge base.

But, a support center should be careful with how a rep communicates with customers. Unless you append e-mail messages with disclaimers to the contrary, every e-mail or chat message to customers comes from your company, not just from a rep. Because on-line correspondence can have legal significance, some support centers take the precaution of requiring supervisors to review e-mail and chat messages before reps send them.

Downsides to Live Support

Costs are the key downsides to relying on reps to respond to all customers' questions. As an example, a May 2001 study by research firm Gartner Group (Stamford, CT), "Contact Center Self-Service Costs," noted that the average cost for a transaction involving a rep was $5 per e-mail message, $5.50 per call and $7 per text chat session. Automated transactions, the report found, average 24 cents on-line and 45 cents by phone if callers only receive information through IVR systems.

Bo Wandell, president and founder of service bureau SafeHarbor Technology (Satsop, WA), says more clients and companies are migrating to automated Web-based support.

Wandell cites a December 1999 paper from Forrester Research (Cambridge, MA), "Tier Zero Customer Support," which shows that when a rep answers a business-to-business support request by phone, the average cost is $33, compared with an average cost of $1.17 per support request when customers can locate answers within an on-line knowledge base.

One of SafeHarbor's clients, a large multinational high-tech firm, now handles between 70% and 90% of all requests for support by directing customers to a knowledge base on its Web site. Before the client signed on with SafeHarbor, that percentage was 30%.

Laying aside the issue of costs, some questions are easier to answer by phone than on a computer.

"There are many technical issues that are more difficult to solve by e-mail than by voice, such as how to configure a computer to connect to an ISP (Internet Service Provider) and receive e-mail," says Chuck Sykes, senior vice president of marketing and global alliances for Sykes Enterprises (Tampa, FL), a global customer support outsourcer.

Live on-line support can be more costly than support by phone, especially if the first rep a customer communicates with doesn't have a sufficient answer. The cost of responding to e-mail varies radically, from $2.50 to as high as $40 per transaction, according to the May 2001 Gartner report.

But, as Sykes cautions, don't focus solely on per-transaction costs. If it takes 12 e-mail messages to solve the same problem that a rep could otherwise fix in one call, it doesn't matter that e-mail is nominally cheaper.

Sykes says, "In theory, when the phones are not ringing, agents can answer several e-mails at once." But, even when a center provides its reps with a collection of templates of responses to save them the time of composing their own, it doesn't pay to wait until support reps aren't busy handling calls to direct e-mail to them. Many support centers find that they're more efficient when they set aside time for reps to focus on answering e-mail.

Sykes has more to say on this subject. "When the templates do not precisely meet the customer's issue, their benefits are cancelled out when the customer sends back an e-mail demanding a more detailed response, or picking up the phone and calling in anyway."

Automated Support

There are many questions with consistent and straightforward answers for which automated support works just fine. It makes good business sense to enable your customers to call your IVR system or visit your Web site where they can get a quick answer to these common questions.

Companies are finding that they can enhance support by setting up on-line knowledge bases where their customers can view and download the information they need, rather than scribbling notes as they speak to live reps. Knowledge bases free up reps to handle questions that require their skills.

Of all the automated options, IVR systems have been around the longest. According to Bruce Pollock, a consultant who works with IVRs (including speech-rec-supported platforms), support centers frequently use IVR systems to identify customers or the products for which they request support. Technical support centers also refer customers to IVR systems to schedule appointments for repairs or to hear how to perform certain procedures like connecting to the Internet.

Some support centers send automated responses to customers' e-mail messages, plus links to helpful Web sites. Automated e-mail systems can also notify customers of problems, like flaws with products and ways to remedy them.

The advantages of automated support, besides cost savings, are immediate answers. Since the late 1990s, more support centers have introduced technologies like speech recognition and knowledge bases that learn over time, all of which make

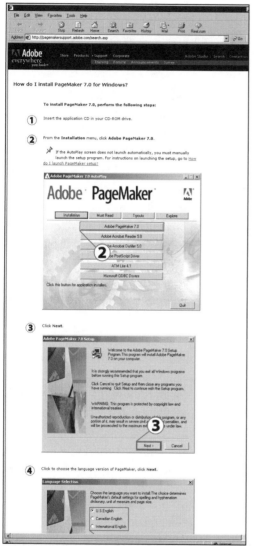

Excellent Web self-service tools often make problem-solving easier than picking up the phone and getting instructions and the solutions from the reps, usually after a 10-minute-but-seems-like-eternity hold. Customers can always refer back to the screens instead of scratching their heads trying to remember the jumble of instructions from the reps or flipping back to scrawled notes on coffee-stained paper.

automated support easier for customers to use and to retrieve the information they're searching for.

You're likely to hear different claims for systems that automate support. For example, Island Data (Carlsbad, CA), which hosts software that generates automated responses to customers' e-mail messages, says its Express Response service lowers operating costs between 25% and 44% by reducing the number of support reps companies need on staff.

But, remember that quality support ultimately depends on the expertise of the reps. Without them, neither your customers nor your automated systems have any place to go for information. Automated support is an adjunct, not a replacement, for live support.

Allow customers the opportunity to reach live reps, whether it's pressing zero while listening to information from your IVR system or starting a live text chat session with a rep from your Web site.

Downsides of Automated Support

Automated support has its limitations. It does not give the reassurance to frustrated customers that properly trained live reps can provide. According to Sykes, many automated systems neither ask customers if the customers' questions have been adequately answered, nor do they record automated transactions.

Each method of automated support has its weaknesses. IVR systems are best for answering questions with a limited set of possible answers. But they're less helpful for illustrating answers to questions that require detailed explanations.

On-line knowledge bases are utterly useless for customers who have problems connecting to the Internet. Telus, the local phone company serving the Canadian provinces of Alberta and British Columbia, is among the savvy companies that can send answers to frequently-asked questions by fax. But that still leaves customers without fax machines out of luck if the only source for answers is an on-line knowledge base.

"E-mail and Web self-service can be very expensive if you don't understand the process and set it up right," says James Mills, director of product marketing for outsourcer TeleTech (Denver, CO). "If the responses they deliver are inadequate, they will generate more replies, by agents and customers, that will prompt phone calls and escalations to senior agents and staff."

Also, customers may become frustrated because they end up spending more time searching for information than they would have spent waiting on hold to speak with a knowledgeable support rep.

To use an on-line knowledge base requires some knowledge on how to use it. Co-author Brendan Read's wife, Christine, works on the Web team for a large company based in New York City. She, her colleagues and members of the on-line customer service team receive not only inquiries from members who are confused about how to use the Web site, but also suggestions on how to improve it.

Chuck Sykes agrees. "Generally most people are really not that tech-literate when it comes to solving problems on their own with self-service," he says.

The technology for automating support is also not cheap. It can cost upwards of hundreds of thousands of dollars to millions of dollars in hardware, software installation and integration costs, depending on the number of support requests you receive. And that technology is still only as good as the knowledge base that supports it.

You might find support to be more costly and the per-incident costs much higher with self-service than you had expected. Fred Van Bennekom, principal, Great Brook Consulting (Bolton, MA) doubts whether offering automated support diverts sufficiently large numbers of calls to merit its value strictly in terms of cost savings. Rather, he suggests, automated service, if it's available on-line, can induce Web traffic that did not exist before.

Remote Servicing; Its Benefits and Weaknesses

One customer contact method unique to support centers, and which is extensively used by internal support centers, is remote servicing. This is where the product or service suppliers' reps investigate, diagnose and solve the customers' problems by using technology to get inside the problematic device; then, when necessary, downloading and installing fixes. The reps can also suggest new software or network services and electronically sell and deliver them.

Remote servicing can result in faster problem resolution and greater satisfaction. According to Kevin Wilzbach, director, integrated contact center services for Convergys (Cincinnati, OH), a large global customer service/sales and support outsourcer, technical problems aren't always readily apparent and are sometimes the result of unexpected compatibility issues. Through the use of remote servicing technology, companies can 'make' it happen rather than hoping the customer will get it right, eliminating possible frustration.

For Convergys' clients and prospects, remote servicing reduces costs by helping increase first call resolution percentages and by reducing product shipping costs. In addition, it can help generate revenue by offering reps an opportunity to sell and push out on the spot if an upgrade or complementary product may be the suggested fix.

But remote servicing has its limitations. Sykes says remote servicing hasn't reached its full potential because of the customers' privacy concerns, i.e. that the company and others can read their data. Also, there has been so much hardware and software built-to-order customization in the hardware and software arena that the method is not always practical.

"For this method to work companies need assure to customers they won't look at or take their data," says Sykes. "They also need to simplify their customizations, and some computers have over 10,000 customizable components, to a more realistic number, like 400, to take care of remote servicing."

Bill Rose, founder and executive director, Service and Support Professionals Association (SSPA; San Diego, CA) recognizes the strengths and weaknesses of remote servicing. "The pros of remote servicing are that you can collect information that can, for example, warn customers of a system failure before it happens, such as if they are using too much bandwidth," he explains. "The cons are that you are tapping into their machine when they are not aware. That can cause problems with firewalls and with other applications they have."

Paying Support Costs

As we noted earlier, support is not inexpensive. Experts calculate it to be anywhere from 5% to 25% of corporate revenues. Al Hahn, president of Hahn Consulting (Portland, OR), estimates that the support costs for packaged consumer hardware or software often begin at $149 per unit, equal to or more than the cost per unit to develop and sell the product.

The big question many companies face, especially those that sell to consumers, is how to recoup their support costs. Should they include support in the price, not incorporate the costs in the price or charge separately?

One option firms can consider is requiring customers to pay for support calls rather than offering toll-free support hotlines. But the savings to companies that eliminate toll-free support are small, given that telecom costs typically accounts for only 10% or less of any support center's operating expenses. And whatever these companies save in telecom costs, they could lose in revenues to competitors that offer toll-free support.

Another option is to absorb the costs of support, while at the same time keeping prices down to compete for market share. The downside is that the support center is less likely to generate revenue and reinforces a common perception that the support center is merely a corporate expense.

The third option, incorporating the costs of support within the price of products or services, risks opening the door to competitors that can afford to absorb the support costs and charge lower prices.

Cable, DSL and Internet Service Providers use incorporated support that they advertise as toll-free. When one firm, such as AOL raised its rates, others followed.

The fourth option, charging for support, or fee-based support keeps list prices low, shows customers exactly what they're paying for, but risks creating the perception of 'nickel-and-diming.' Banks have long maintained this practice, which has led to some legislators calling for abolishment of ATM fees and other such charges.

The advantage of fee-based support is that it enables your support center to generate profits for your company, if the customer base is willing to pay. Many companies, such as Oracle, have come to depend on support as a revenue stream.

Three Ways to Provide Customer Support

The first method is indirect. The customers contact customer service first, and if they have a product problem, the call is escalated to support. This choice works best with durable and/or comparatively nonessential products and services, such as small household appliances and consumer electronics where only a small percentage of calls require fixes.

Second is to have separate customer support numbers for centers. This option applies to products and services, which break down more often and/or are essential, such as computer hardware/software and communications (phones, Internet services and on-line commerce).

But in all cases you must connect customer service with support, and vice-versa. For example, telcos had allocated the 611 exchange for problems, and other numbers for service. The same goes for banks and credit card issuers. They have separate personnel who handle phone or on-line bank transactions, and others who fix on-line banking problems, such as missing passwords.

Many high-tech firms such as ISPs usually have separate customer service and support phone numbers. If customers have accounts or billing problems, they call customer service. If they have e-mail or Internet access problems, they call support. If customers call in with billing problems but say they also have technical problems, the customer service agent passes these questions to tech support.

Third is self-service. But for this connection method to work it must have e-mail addresses, call me buttons, links on Web sites, clearly indicated phone numbers and 'zero out' options to live reps.

We, and the experts we interviewed for this book and for our magazine articles on customer support and service cannot stress the importance of live agent and rep contact enough. Your customers are people. If people need to reach people but cannot get them, chances are they will not be your customers or be very loyal customers for long.

"What typically happens is that a company launches a new product or service that is innovative, grabs great market share and prices it so that it has a high enough profit margin to absorb support," explains Hahn. "Then as more companies enter the market prices come down, which squeezes support costs. Eventually they are forced to break out services like delivery and support with separate charges."

Al Hahn is a strong advocate of fee-based support. He also points out that charging for support helps companies control their support costs by encouraging customers to solve problems themselves or to use automated systems.

"By charging for support you educate your customers to understand that it isn't free, that there is a cost to it," explains Hahn. "But you also obtain buy-in from top management for your support center because they see it generating profits."

There are several ways to charge for support. They include pre-paid support by the hour, or having an hourly rate based on customized report levels such as for self-help, queued rep and dedicated rep, with each upgrade costing more. This latter method can be marketed as 'gold,' 'silver' and 'bronze' service. But, pre-paid support also can be charged by level, e.g. level-zero, level-one, level-two. Many leading high-tech firms like Sun Microsystems use this method.

Customers and companies arrange these in service level agreements (SLAs). The SLAs state what level of support the customers get for their money. Frequently, and depending on the product they include field service.

Another pre-paid support method is per-incident, usually paid for by a credit card but sometimes invoiced. This can take many scenarios, with the typical ones being a customer calling or e-mailing in, being told how much the support will cost, and supplying their credit card number before being connected to a rep.

Still another method is a service plan, typically offered by cable companies and telcos. You can be responsible for your own internal wiring, from the service to the idiot box or to the chatterbox. These plans are usually less expensive than paying per-incident. In any of these instances, the cable or phone company continues to maintain and repair phone service to your service entrance.

Fee-based methods come and go, as technology, corporate practice and customer acceptance of them changes. For example, as SSPA's Bill Rose explains, the 900 line method became popular in the 1990s but soon faded away. Among the reasons: companies began putting blocks on 900 numbers as they are also used by dial-a-porn services, and customers found they were paying for metered time when they didn't have to. "Sometimes a customer called in on the 900 line, was put in queue and then was told by the rep that they didn't need to call in because the problem was caused by a product defect," recalls Rose. "But the support organization had no way of refunding them."

Should You Charge for Support?

Fee-based support has long been the practice for manufacturers and distributors of appliances like washing machines, and vehicles, industrial equipment and computer hardware. These goods break down and all wear out, so the owners or operators either fix them or replace them.

Fee-based support also has a history in the higher-end, high-cost business-to-business software marketplace because these customers have long understood the rationale for breaking out the costs of items. They want to know exactly what they're paying for.

In turn, paying for support encourages the owners and their employees to take good care of such products and services in order to keep their costs down. It is always cheaper and more convenient to change the fluids in a car at the recommended intervals than to fix or replace a cooked engine or transmission in the middle of nowhere, where 'Jerry's Junkyard' eyes such owners almost as hungrily as the mosquitoes in the nearby swamp.

But if your firm sells technology — hardware, software and services — to consumers, asking them to pay for something that has long been free might be difficult, but it's not impossible; you just have to examine your market very carefully.

As a general rule, consumer computer equipment manufacturers, Internet service providers and cable and phone companies do not charge for support, with the exception of on-premise wiring. Software companies have more leeway. Some charge, some do not, but the trend to toward fee-based support service.

For example, Educational Programs and Software (better known as EPES) provides software applications for school administrative functions. It offers no charge for support if the customer has had the program less than one year or has entered into a support contract. Otherwise, the company charges for support on a charge per incident basis.

Then there's Trigram Software, which sells its software, AcuBase Pro, with three hours of free customer support for 90 days from the date of purchase. After that, the minimum charge for support is $15 for the first ten minutes, then $1.50 per minute after the first ten minutes. After that, one hour of support is $75; three hours is $198. The support fees are charged to the customer's credit card of choice. This is just the tip of the iceberg, there are many more companies similarly charging for support service.

Fred Van Bennekom explains that the reason hardware, ISP and telco vendors don't charge for support is because they regard support as a service device to keep their customers loyal, although there are exceptions.

For hardware firms and telcos, there are fewer calls that chew up support costs compared with software, especially consumer software. It is also much less expensive for a manufacturer or telco to fix a problem over the phone than it is to send someone out to fix a computer, or have the customer bring it or ship it to a repair depot or fix the line.

"Small hardware or software companies do not charge for support because it is not part of their practice and because they need to stay competitive with larger firms," explains Van Bennekom. "But where a firm has sufficient market strength, like a Microsoft, to charge for support, then a fee-based support strategy is very effective at diverting calls to self-service. When companies move from free to fee they typically leave the Web-based self-service as the free option."

For instance, with Microsoft's Windows and Office products, the customer gets two free calls or e-mail inquiries (or one of each) to tech support, each subsequent inquiry costs around $35.

Hahn asserts that consumers who purchase products like desktop computers may be willing to pay for support. That's because customers differentiate between buying a product and buying additional support as an add-on after the warranty expires. As with cars, when repairs become the customers' responsibility, customers have no qualms about paying for necessary repairs and service.

The manufacturers have taken note. Compaq's Consumer Products division, Apple and others provide free support during the warranty period, but after that, each charges about $35 per incident or charges the customer a per-minute charge.

"There has long been resistance by companies selling desktop hardware and software to adopt fee-based support because they are worried that if they broke out and added those fees that they will lose market share from customers who don't like paying 'added' costs," explains Hahn. "But what we found when we devised a fee-based support system for our clients was that they didn't lose market share."

Charging for support for other products, like consumer software and Internet services, becomes more challenging because the cost of support is often as high or higher than the cost of the product. But Hahn expects that these companies eventually will have to charge for live support to stay competitive and profitable.

And consumers will pay for such fees, even though they don't like them. One example is the ATM and checking charges at banks. The companies argue that they need the fees to pay for the costs of providing these services.

Another example is the shipping and handling costs on phone and on-line orders.

Balancing Live Agent and On-line Self-Service Support

On-line self-service and e-mail support can help customers and reduce support costs but only if these methods are made customer-friendly and are integrated with live-rep response. This includes training reps on how to provide support and how to use the on-line tools.

That's the message from Rick Kilton, president, RWK Enterprises (Lyons, CO), a consulting firm.

He discussed Web versus 'human support' in a presentation he gave to the Association for Services Management International in November 2001.

He pointed out that Web-based support offers several important advantages. They include:

- Lower support costs, if more customers go to Web you save money.
- 7x24 support for those who don't run 24x7 support centers.
- Capture of customer information, such as who is logging in and when, what services are being used and what parts of the site are being visited.
- Emergency information.
- Support of customers who do not like talking to people, because they may have had bad experiences in the past.
- Can act as an e-mail portal, with formatted messages enabling quicker, less costly support.

But on-line support has these pitfalls:

- Customers can't find information they're looking for because:

One of the long-held 'secrets' of direct marketing is that much of and often the only profit lies hidden in the shipping and handling costs, which not even the most intrepid customer can break out accurately to determine how much profit the vendor is actually making on their order. That's why many products and services are offered 'free', 'plus $4.95 shipping and handling'.

They also make money from selling your file to other direct marketers, unless 'void where prohibited.' Encroaching privacy, telemarketing and e-mail

- The Web site takes them to dead ends.
- The Web site is too difficult for customers to navigate.
- Customers have no other source of information other than the Web site.
- The Web site is too obtrusive (asks too many questions, invades privacy).
- Before letting customers in, the Web site:
  - Can't tell if a customer is mad or happy.
  - Can't tell if or why a customer left.
  - Is designed by 'techies' who don't or can't relate to customers.
- The Web site is not current.

"There are several myths with Web support," Kilton points out, "Contrary to claims by vendors, they do not create loyalty but can enhance it or destroy it. Also Web sites do not necessarily save money by handling basic calls. Instead they may drive away customers if the on-line support is of poor quality and may result in additional phone calls. The customers will see self-service as a obstacle to solving their problems: 'the monster in front of me that's not helping me do my job.'"

On-line correspondence allows customers to reach out to support reps quickly, but compared to phone conversations, it's not always as effective in enabling customers to communicate how they feel and what they mean. Because this information is missed it leads to e-mail 'ping-pong' between rep and customer that takes hours and days to solve. That also leads to angry, dissatisfied and ultimately 'ex-customers.'

But in either case, companies that provide on-line support "don't know when a customer will become angry," says Kilton. "They also don't know when they become mad enough to leave. Ninety percent of customers don't 'anti-spam' laws may make that practice eventually go the way of the door-to-door salesman.

Hahn suggests a couple of fee-based support strategies. One, you offer top-quality user-friendly self-service, then charge for level-one or level-two support. Or two, provide free self-service and level-one support but you identify your largest customers, get them to register with you, and offer them a no-queue access to level-two for a fee. "These buyers are the most sophisticated and tech

---

let companies know they're taking their business elsewhere."

To solve these problems and to keep customers Kilton suggests taking the following steps:

- Put in very clear paths to help on the Web site and in e-mail messages, including phone numbers where customers can call for help.

- Besides developing support reps' problem-solving skills, train reps to acknowledge customers' feelings.

- Save the path the customer took on Web or IVR to avoid having the customer repeat it. This helps reps solve problems because they can see where the customers were on the self-service. Reps could also then advise customers how they could have solved the problems, saving them time and hassle. That way the customers will successfully use the Web site or IVR the next time without calling.

- Walk customers through the Web site and teach them how to use it.

- Teach support personnel how to use the Web site. Many of them do not know how to use the Web-based material for their own troubleshooting.

- Measure customer satisfaction with Web support, or if logged in, send an e-mail link (people don't like long questionnaires).

- Provide e-mail addresses from where visitors to your Web site can easily contact support reps to ask a few questions. If a customer makes comments, have the reps place these remarks in your call tracking system.

- Ask a variety of people, preferably nontechies from among your friends and family, to try out your Web site. Their comments and suggestions are likely to mirror those of your customers.

"There are no 'dumb customers,' if you want to make money from customers no matter what they do," Kilton points out.

savvy and will have naturally gone through your self-service and want to bypass level-one," says Hahn. "They will not mind spending money, say $12 or $24 to get instant expert support."

If you do charge for support, make sure you have top-quality reps on hand to take the calls. Also, limit queues or offer the choice of queued support at one price and instant connections at a higher price.

Because you're offering customers an intangible, you should provide a tangible, like a paper and bound report in a nice looking box or a maintenance upgrade.

If charging for support is not practical, Hahn recommends that you send to customers no-charge invoices for the list price of the service to show the value of what they're receiving. "You'd be surprised how impressive a $5,000 discount looks on paper," he says.

And with some products, like software, customers who receive them for free as downloads would be willing to pay for support. In some circles it is no longer cool to *buy* software. While software can easily be copied, the same can't be said for live-rep support, especially at the upper levels.

Additional Operations

Because call centers work closely with other areas of your company, it helps to be knowledgeable about the impact of your support operation on your e-commerce and fulfillment efforts.

Three excellent sources of information we recommend are A Practical Guide to CRM Logistics and Fulfillment for e-business, and The Complete E-Commerce Book all by Janice Reynolds.

Four

CHAPTER FOUR

PROVIDING
LIVE SUPPORT

Providing Live Support

C H A P T E R **Four**

There are three realities all support centers should be built upon.

One: At some point your customers will want to speak with a live human being to solve their problem.

Two: Contacts come in two rough sizes, basic and expert (with variations in between).

Three: When all else fails the issue ends up at the manufacturers'/ providers' doorsteps, i.e., engineering/product development, depot repair and field service and support.

They made it. They fix it. Or they face the rath of customers, customer advocates, their lawyers, and if the issue and/or vendor is big enough like software holes that computers viruses can infest and mutate in, by the press and politicians.

Support Levels

A support center can save a lot of money, and keep its expert, well-paid support reps happier in problem solving by having the basic calls answered by the lowest-trained-and-paid staff. A basic call is one where there is a simple problem that could be resolved with a simple, easily remembered solution like 'did you plug it in' or by looking the matter up on-line or in a manual. An expert call is one where the equipment is FUBARed ('fouled' up beyond all recognition) and the assistance of someone who really knows how it works is needed to fix it.

To help organize their customer support, companies usually divide their reps into formal levels or tiers, although there is no industry-wide agreement on how many levels there are or should be, or what they can accomplish. Consultants leave the number and variety of the levels or tiers up to the individual centers to determine because every company and product/service is different. If you make sophisticated hardware and software and sell them to other companies, governments and professionals, chances are the persons calling your company are as knowledgeable if not more so than any of your support reps.

So here is a rough guide:

Level-one is basic problem solving. Guiding the customer through a simple installation process, i.e. customer handholding is a level-one task. Reps diagnose the problems' symptoms with the assistance of a knowledge base and propose solutions to them that the customers then try out. Rarely do they go into the specifics of, for example, computer hardware or software.

Everybody needs a level-one rep sometime. Even a basic or expert rep who comes home, finds that they can't get the #$%^&* thousand-dollar dust collector to work even after reading the !#$%^&* manual and drilling through the !#$%^&* Web site and IVR menu.

Level-one support is often referred to in the support industry as 'read the (friendly) manual' (RTM). That's because many of the answers lie in the manual, which customers, as often alleged by support reps, are too lazy to read.

But as anyone who has gone through a manual knows, too many manuals are anything but 'friendly.' Many manuals can only be understood by members of the same species of geeks who wrote them.

Moreover, there are many instances where customers do not have the manual at hand; or are too pressed or stressed to obtain a copy of the manual; or it's inconvenient to use a manual. Let's say, for example, that your company sells software or laptops and a customer is calling your support center from a wireless phone while waiting to catch the 8:55 to Phoenix for a business meeting. Also assume it's 20 minutes before takeoff and they can't open their presentation program. You can't expect that individual to have overstuffed their briefcase with the manuals beforehand or to patiently drill through the self-help.

That's why you need a person at level-one. Level-one requires reps who have the ability to ask the right questions, and who are logical and can problem solve and yet who can empathize with the customers. They are the front line of your company.

What happens at level-one makes or breaks the relationship. A customer who doesn't like the answers or the treatment will walk away. According to consultants like Rick Kilton, president, RWK Enterprises (Lyons, CO), 90% of customers who do say "Adios," do not tell the offending company that they're leaving.

> **WARNING:** If you make sophisticated gear and services and have equally expert customers, your level-one support reps have to be as smart and as well-trained as the callers. Your customers paid good money to buy your stuff and they don't want to waste their time going through dumb step-by-step problem solving.

Level-two is complex problem solving. These reps deal with more detailed diagnostics and virtual surgery, like getting into the customers' machinery and downloading bug fixes. Level-two inquiries can also be referred to as 'chuck the (friendly) manual' because they often require the support reps to devise their own solutions and try them out on the customers.

Level-two requires reps who have computer/technical knowledge, in some cases computer science or electrical engineering degrees. Consequently, they are paid more than level-one reps. A level-one rep could earn $22,000/$23,000/year depending on the labor market where the center is located but a level-two rep will expect $35,000/year and up.

Depending on the product and application, experts say more than half of support calls — as high as 80% are level-one and no less than 20% are level-two. They also advise that if you give free support you risk watching your support costs heat the ionosphere up because seemingly everyone will be dialing in, instead of practicing RTM. But if you charge for it, like Microsoft does, then expect support volumes to be lower.

Level-three and level-four are engineering and product development. Level-three calls are diagnostic. They often involve reps logging into the customers' computers and fixing them. Not all companies have level-three and level-four support. As a general rule the more complex the product or service and the larger the company the more the number of levels you need to have and staff.

Level-zero can be considered the 'receptionist' level. A support call comes in and the attendant, human or electronic, directs it to the right support person, account representative or field service, depending on how the company is structured and their level of automation.

Although still around, level-zero, in general, faded away with the growth of

IVR/autoattendant backed by sophisticated call routing systems. It became purely a call logging function that involves no intelligence applied to solving the problems.

Fred Van Bennekom recalls that back in computing's Bronze Age from the mid-1970s to the mid-1980s, companies like Digital would have pools of clerks taking level-zero calls.

"These clerks would ask the customers who they were, identify their problem and then either forward the matter to the support rep designated to handle that problem, or have that support or field service rep call them back," explains Van Bennekom.

But level-zero has come back with fee-based support. Some companies now supply free 'Level 0.5' where reps provide some basic support before switching the call.

Divide By Channel

Another new and increasingly popular way to divide support is by channel. Level-one becomes e-mail response only, or as an opt-out of a long queue, and level-two takes voice inquiries or escalated e-mails.

With channel division, customers' e-mails are transmitted to a support center either in-house but they're often sent to a service bureau in a developing nation like India or The Philippines or to a bureau or an in-house center in a developed nation like Australia, Barbados, Canada, Costa Rica, New Zealand or a province like Northern Ireland that has low costs. If you have an excellent contact management system you can have your customers' trouble tickets acted upon and updated anywhere in the world.

This method takes advantage of low per-transaction costs and less expensive highly qualified labor in other nations. Companies can cut operating costs by as much as 50% by locating or outsourcing to support centers outside of the US (see chapter 6).

Division by channel allows a company to get around the accent and cultural issues. Many Indian agents and reps, taught British English, cannot talk effectively to Americans, hindered by accent and by fast rates of speech. E-mail eliminates many but not all of these problems and e-mail allows supervisors to edit an outbound e-mail for accuracy and language usage, and bring recurring problems back to reps for corrections.

✦ **NOTE:** Some companies are now teaching their reps to understand and speak in American and a few are going over the top by allowing and encouraging their agents and reps to fake that they are Americans: a practice that risks receiving hostile reaction from American consumers.

The Downside

Your customers may not get immediate response to productivity-hampering problems. E-mail support response times are measured by hours, not by minutes. There is no real-time dialogue unless you set up a chat session, which adds to the costs — to the point where voice is cheaper.

Also you must carefully supervise and train your agents on e-mail response to ensure accuracy and language use. And for many level-one RTM calls, customers want that human voice that reassures them. They can't get that warm and fuzzy feeling off an e-mail.

Support by channel requires a very robust contact management system, thoroughly integrated at all support centers, updated religiously. The data network and carriage must be flawless. The level-two reps must be told who handled the e-mail at level-one, and track it. Some companies, like Convergys, house and run their routing, switching and data handling from the US, with their support desks in other countries like India acting like dumb terminals.

Escalation Issues

The challenge support centers face is balancing costs and service. Escalating trouble tickets is not cheap. The more levels you have the more you must spend in staffing and training for those levels, which become increasingly expensive the higher you go. While a level-one support call can cost $20-$35 to handle, costs take a radical jump to $100 when a level-two rep is brought into the mix.

One way support centers can save money is by implementing practices that increase first call resolution. (See the Primavera Case Study in chapter 2 and the discussion on support center practices in chapter 12.) Such practices also make life easier for upper-level agents and shorten handling times by diagnosing, handling and dispensing with simple issues, just as you would see a nurse or your doctor for an ailment before being referred to a specialist.

But the tradeoff — requiring level-one reps to have better qualifications and training them longer — is higher agent compensation and staffing cost.

Field Service

Another area that involves live support is field service and despite a rise in the adoption of sophisticated centers dedicated to customer support, there will always be a need for field service and support depots: the traditional outlets for fixing things.

If your company supports products, especially tangible items like toasters or chairs, it's likely that your support center will need to work closely with warehouses, repair depots and field service organizations.

While support by phone and on-line diagnostics are, in many circumstances, displacing field service, it is still the fastest and the most convenient way to solve problems with physical objects like cars, refrigerators and computers. (With services like roadside assistance, sometimes it's the only method for getting support.)

There are also malfunctions that cannot be fixed by a support rep: fans, power supplies, television sets and more. "Screwdrivering" around computers, for example, can even violate a customer's warranty. Nor are support reps or your customers responsible for what lies outside their building or department, like the wonky ADSL modem at the local telco central office (for more about DSL check out Janice Reynold's book <u>A Practical Guide to DSL</u>) or the rodents that are hooked on the taste of underground coaxial cable.

For example, Dell is famous for its support, which includes field service. The computer maker contracts with BancTec in the US for field service. But, the service contract requires customers to call for support first, and try and solve the problem with the rep before Dell dispatches a technician from BancTec.

The Loaner Conundrum

Field service can fix your customers' problems there and then. This rescues your customers' productivity compared to the downtime when the appliance, machine or vehicle is at the shop, or working with "loaners" that in most cases are seemingly designed to be not as good as the equipment or machine being serviced. And, then there's the issue with computers. If the product being serviced is a computer, the hard drive needs to be loaded (hopefully correctly) with software, applications and connections, to enable customers to use it.

It is amazing how much time one wastes just getting used to a different computer or car, even one with the exact or even better performance characteristics. And if the device or vehicle is out long enough, how long it takes to get reacquainted with your original gear.

Our intrepid co-author Brendan Read speaks from experience (which is also why he's co-writing this book). When his old Librex 286 laptop died while he was living in the UK many years ago [he used to get strange stares working the thing on trains, such was the novelty] the machine had the good sense to crap out just before the warranty expired.

Read shipped the computer from England to his parents in the US who then shipped it to a contract repair depot; there were none in Britain. The depot eventually shipped it back. In the meantime he had to rent an electric type-writer, which forced him to go back to the messy old days of bonded erasable paper, carbons and whiteout. Fortunately he is old and decrepit enough to remember how to type manually (in case the electric crapped out) in true jour-nalism style with two or three fingers, and rely on his biochemical spell check and Oxford dictionary.

Speaking of intrepids, one of the most, (ahem) 'enterprising' examples of depot repair Read has seen and utilized is in Ravena, NY. Practically next door to Marshall's Garage, the local Chrysler dealership, is an Enterprise car-rental storefront and a Dunkin Donuts. When Read's 1993 Intrepid's transmission blew its last gear on a trip to the Catskills — not wanting to fry the tranny's limp-in mode on the bumpy, rutted, debris-strewn demolition derby racetracks that pass for roads in the New York/New Jersey area — he took the car to the Ravena dealership. He then went next door and rented a Chevy Cavalier from Enterprise.

"It took a while for me and my wife to get used to the Cavalier, which is much smaller than the Intrepid," says Read. "But when I returned the vehicle, to pick up mine, I found that I had to get used to driving my own car again."

Arranging for a field service consultant is often the best and only way to fix computer software problems that interfere with other software, network appli-cations and Internet connections. (See the Brendan Read story in chapter 1.) These on-site gurus may be the only surgeons who can restore and revive a machine after customers erroneously open a virus-infested document.

Five

CHAPTER FIVE

SUPPORT CENTER
ADJUNCTS

Support Center Adjuncts

CHAPTER Five

Besides employing support reps on your premises, you have the choice of allowing reps to telework (i.e. work from their homes), bringing in additional support reps with the help of outsourcers who furnish you with staff from their locations, or using insourcers who provide you with staff at your locations.

In this section, we describe how teleworking, outsourcing and insourcing can serve as cost-effective alternatives or adjuncts to a support center.

Teleworking

Support reps generally have the personality traits necessary to effectively work from their homes, including the self-discipline to answer customers' questions on their own. Support reps typically take ownership of customers' problems, and they ask others for assistance as a last resort — often only when they're about to escalate a customer's request for help to a higher-level rep.

Cost savings are among the clearest benefits of employing teleworkers to support customers.

According to consultant John Heacock, who works from his home in Parker, CO, companies that allow reps to telework reduce their facility expenses by about two thirds.

One reason is that you can keep your support center smaller if you don't have to make room for every rep. You can cut the fully allocated cost of supporting each employee, with workstations, computers, phones and cable, from $35,000 to as low as $10,000.

An equally important benefit of allowing support reps to work from their homes is that it presents your company with another way to attract and retain this group of employees. You're able to bring in qualified workers who otherwise would be unavailable to you since they cannot easily travel far from their homes because, for example, they are parents with young children, caretakers of elderly parents and people with impaired mobility. And, if an experienced rep decides to move, teleworking allows that valued employee to continue working for you wherever they are.

By employing teleworkers, you avoid having to locate your center in the heart of communities where rents and salaries are high, like Silicon Valley and the Route 128 electronics belt around Boston. Also, some people are willing to work for companies that pay what appear to be lower salaries if the option of teleworking spares them the hassles of commuting.

That's not the only benefit employees gain from teleworking. Employees also save money. Fewer trips to the gas station, or maybe you don't need that rustbucket and the insurance ransom to keep it on the road, and no more work shirts, makeup, hairspray and pantyhose. Heacock says these savings can be as much as $4,000 per year.

Equally important teleworking gives reps a valuable perk no money can buy: *time*. The 'tele-commute' gives the reps back hours that they heretofore wasted in transit, or stuck on the road in their gas-sucking monster beater, fouling the air and melting the icecaps. Time for family, to have a life, and time to make the passing of every grain of sand in the hourglass to count for something.

Teleworking confers additional benefits to companies besides cost savings and larger labor pools from which to select support reps. Heacock notes that some companies have increased productivity at their support centers by between 40% and 80% when reps telework. There are also reports that turnover drops to near zero when an employer allows support reps to work from their homes.

Tom Pugh, who is with telecommunications equipment manufacturer Siemens, reports that during 1999 and 2000, his firm successfully enabled 200 support reps (who assist Siemens' employees and customers) to work from their homes instead of having to commute to four support centers throughout the US.

"Having the support reps work from home has worked out very well," says Pugh. "It has cut down on attrition and it has allowed Siemens to more efficiently use their buildings, like for other offices and training."

Chances are, therefore, that if you offer teleworking, either from the start or for reps who prove themselves, they'll jump at it. They'll soon perceive that working in a building is only for 'newbies' or for reps who need hand holding. The 'true brethren' will work at home and communicate with each other through text chat and instant messaging.

You also expand your reach of workers in a given metropolitan area by offering teleworking.

They can work for your fantastic organization, notwithstanding the fact that it is located in a 100-mile sprawling congested, pollution-riddled mess of malls of bankers boxes that necessitates being stuck on I-whatever at 7 am, exhausting their fuel and their good tempers.

Or perhaps your small business could be next to the boss's favorite cigar vendor in Greenwich, CT, and your reps live in Greenwich Village. With teleworking, they don't have to bounce over New York City's notoriously pothole-riddled Bruckner Expressway or play rat race on the Connecticut Turnpike jammed with lottery ticket buyers. Or push against the suited lemmings crowds in Grand Central to catch the reverse commute train on Metro North.

Also, you may find teleworking a viable option to fill undesirable evening and weekend shifts. Many people do not like to commute at night, when the risks of crime and traffic accidents go up.

Chances are reps don't want to crawl into the night to do the ghoul shift, grabbing a ratty cup of grounds at the all-nite buy-and-burp on the way and watching out for denizens of darkness when parking their car. They'd rather be on-line or on the phone from the comforts of their own homes at 1 am, like many of your customers.

Creating the Right Environment

Even if all support reps at your company telework, you still have to provide them with certain resources to do their jobs. For instance, extensions to phone switches so your call routing software can direct customers to remote reps. Teleworkers also need access to essential software, including tools that let them view current information about customers. And you should be able to record and evaluate each rep's communication with your customers, no matter where the reps work.

Chances are that teleworkers who provide technical support will personally own computers that are at least as powerful as what you provide reps in your support center. By definition, technical support reps must know something about machines, and usually they can figure out what works best for them. But don't assume. Whatever technical support reps must have to help customers — specific equipment, high-speed Internet service or remote connections to your corporate network — verify that it's all available from where they live.

Ideally, a rep's home office should be on an equal footing with a rep's workstation at your center. It's vital that you're consistent about how you specify, approve and purchase furniture, software and communications equipment for reps. That includes making sure your company has a clear procedure for choosing and setting up chairs, desks, phones, desktop computers and computer monitors in a way that's ergonomically sound for all reps, whether they're on-site or off-site.

Workstation placement in reps' homes need not be difficult, given that many makers of office furniture design their units to accommodate compact spaces. Co-author Brendan Read knows this from experience. He bought a corner workstation with a manually-adjustable monitor and keyboard tray from Staples because the store offered the only affordable model that would fit his home office. The workstation, which he purchased in 1998, has proven to be adaptable and rugged, even after he took it apart, packaged it and re-assembled it in his new home in Victoria, BC, Canada.

Minimizing Risk

A telework program provides business continuity. Teleworking reduces the likelihood that a disaster at or near your support center will prevent all of your reps from being available to assist customers.

Also, since you decrease headcount within your center by employing teleworkers, you need fewer on-site generators to provide power, heat, air conditioning and lighting. That makes your company less vulnerable to outages if your center is in a region of the US that has to limit how much energy it uses, as occurred in California in 2001.

Enabling reps to telework can also cut workers' compensation costs, which typically account for between 10% and 60% of a firm's total insurance bill. Reports suggests that, following the September 11 attacks in New York City and Washington, DC, workers' compensation expenses may climb for employers with offices that staff more than 100 workers. *The Wall Street Journal* reported on January 9, 2002 that most re-insurers would not cover terrorism. Since insurers, in turn, can't categorize terrorism separately, they may slice their overall workers' compensation coverage.

In addition, Heacock observes, workers' compensation claims from teleworkers are rare. "One reason there hasn't been that many is that teleworking employees don't want to jeopardize the advantages of working from home," he explains.

Teleworking Downsides

If you're considering whether to allow or encourage teleworking for a support center, take a hard look at your company's management philosophy. Must supervisors see employees to be sure they're working? Does your company place enough trust in its employees to allow them to telework? Does it have systems in place to monitor a teleworker's efficiency and performance?

Not every support rep is right for teleworking. Some lack the experience or discipline to work without supervision. Others prefer the camaraderie of being in a workplace. A few reps may not live in homes that are suitable for teleworking, i.e. they may lack the separate space that allows them to work effectively, or they might be surrounded by too many outside distractions, like trucks or trains that frequently rumble by.

Whatever teleworking arrangement you set up, remember: What works is what counts. You may find that the answer is a mix of on-site support reps and teleworkers, where, for example, new and inexperienced reps communicate with customers from a small site that serves as an extended training floor. In this setting, the teleworkers are the experienced reps.

Insufficient bandwidth is another possible downside for the technical support rep who teleworks. These reps must be able to tap into knowledge bases and other software, which can take a long time to load on their computers if they only have dial-up Internet access.

"Telecommuting doesn't really lend itself to tech support, which involves frequent updates and training on new product releases," says Chuck Sykes, president of Sykes Enterprises, a Tampa, FL-based global support outsourcer. "We haven't yet cracked the knowledge management dilemma. The e-learning tools and bandwidth aren't there yet to make remote agent training feasible."

But according to Heacock and other advocates of teleworking, high-speed Internet services are becoming available in more parts of the US, which means faster and easier off-site access to call center software, and support reps don't necessarily have to connect to their companies' networks. Some developers of call centers software can host their products from their own servers or from those of their partners, which means that all reps effectively use software that runs from machines located outside their centers.

Outsourcing

In the context of support, outsourcing is contracting out all or some of your operation to a company where reps communicate with customers on a client company's behalf.

The principal benefit of outsourcing is that it enables you to avoid paying for people, processes and technologies you don't need all the time; thus, at times these costly assets would be idle and unproductive.

The other key benefit is flexibility: the outsourcer can ramp up your program far more quickly than you could add workstations and support centers within your company.

For this reason alone, outsourcing is a very valuable strategy; especially if you experience high seasonal demand for your products, or if you're launching a new product or service. Outsourcing is also effective when a company experiences an unexpected sudden onslaught of customers during a short period of time, which is what can happen, for example, with product recalls.

"Say you're coming out with a new product or version in the next few months," explains Geri Gantman, senior consultant with Oetting and Company (New York, NY). "You know from the moment you release it you're going to get a flood of [support requests] for the next few months. You don't have enough in-house reps to cope.

"The outsourced reps bear the brunt of the wave by answering the simple questions from your knowledge bases and sifting out and escalating those that need attention. When the wave diminishes you cut back on the hours you need from the outsourcer."

Keep in mind that a well-run and well-located in-house support center is as efficient as an outsourcer's operation. The main advantage of outsourcing is lower costs. While these two statements may seem contradictory, according to studies, such as those by the Association of Support Professionals (ASP; Watertown, MA), the median cost of an outsourced transaction is slightly above $15. That's well below median per-transaction expenses among in-house support centers, which report costs that range between $20 and $35. Some companies that sell software to consumers cite median outsourcing costs as low as $5 to $6 per transaction.

"Those costs reflect the nature of the calls outsourcers get, which tend to be lower-level, but also to some degree, a different model outsourcers employ for figuring the cost per call that they charge a client," says ASP executive director Jeffrey Tarter. "When outsourcers take on new clients, they typically charge for incremental marginal costs incurred by taking them on and profits. But they often do not calculate all the costs, like infrastructure and administration, divided by all the calls."

The savings outsourcers pass along to their clients result in part from their economies of scale and their flexibility in selecting sites.

For an outsourcer, the expense of maintaining sites to accommodate 600 seats and to hire the equivalent of 1,200 full-time reps are less than the expense for six companies to set up their own 100-seat centers and each to employ the equivalent of 200 reps. Why? Reps who work for outsourcers often assist customers on behalf of multiple clients. Since each client's support requests peak at different times, outsourcers can manage with fewer reps than each client can by itself.

Outsourcers are not limited by preferences of senior management, like the location of the nearest 19th hole on the golf course. They constantly seek new and affordable locales, outsourcers can more readily locate in lower-cost areas, such as Glace Bay, Nova Scotia in Canada; North Bend, OR; or Bangalore, India.

Many companies, especially well-known large firms, like Compaq, micron-pc.com and Microsoft, outsource their support, as do smaller firms we've profiled in the pages of *Call Center* magazine.

Level-one support lends itself best to outsourcing because a level-one rep needs less training than a higher level rep. A level-one rep's skills with helping customers with one product can apply to helping customers with other products.

Outsourcing Downsides

Fred Van Bennekom, founder of and principal with Great Brook Consulting (Bolton, MA), says outsourcing is fine for level-one calls where customers ask basic and consistent questions.

But he doubts if companies can effectively outsource support beyond level-one. Reps who provide high-level support occasionally have to devise their own answers and draw upon resources outside the support centers.

That's risky in an outsourcing partnership. Support reps outside your company who propose solutions to customers' problems may not be able to receive advice or approval within your company as easily as in-house reps.

Also, support reps outside your company may not exhibit the same degree of commitment to assisting and satisfying customers as reps at your company. The higher the level of support you outsource, the more training you may have to provide to reps who are not your employees.

"There is a high degree of trust that must be established and maintained by the support rep because of the high investment involved," says Van Bennekom. "If a customer knows or finds out that the support is outsourced, they do not look upon the information with the same degree of veracity."

Outsourcers tend not to encourage reps to incorporate the information they gather from supporting customers into their clients' knowledge bases. Nor do they pass on what they learn to product developers who would be able to correct root causes of problems, such as defects, that lead customers to ask for support. Why? Because increasing the scope of a service bureau's responsibilities takes away the key advantage of outsourcing.

Cost reduction is why companies turn to service bureaus, which frequently compete with one another on metrics like average costs per call. Asking reps from outsourcers to contribute to knowledge bases would add to the costs.

For that reason, says Van Bennekom, it is neither in the outsourcers' or the clients' best interest to require outsourcers to find their own answers to customers' questions. When the existing set of answers don't work, outsourcers are better off escalating customers' requests to in-house reps.

Serving Customers from Offshore

To provide better service at lower costs, many companies and outsourcers are shipping their support of US customers to service bureaus in Canada, the Caribbean, Northern Ireland, the Philippines and India. These service bureaus report cost savings of 25% or more while primarily employing college or university graduates, many who are better-educated than American reps. That's because support reps in some countries regard their jobs as middle-class positions.

American service bureaus that focus on support, such as Convergys, PeopleSupport, Stream International and Sykes, also have set up offshore centers. Similarly, overseas service bureaus are seeking support contracts from American firms.

Some outsourcers, such as Stream International, maintain support centers in both India, whose labor costs are half those in the US, and Canada, where labor costs fall in between those in the US and India.

"We grow in both countries in response to clients' requirements," explains Deborah Keeman, Stream International vice president of services marketing. "Some of our clients do not mind having their [customers] handled offshore in India to take advantage of the lower costs. For others, accessibility is an issue, so we accommodate them in Canada."

Given that the written word has no accent, many outsourcers also offer on-line support, which can mitigate potential misunderstandings between overseas reps and American customers. An advantage of on-line correspondence is that companies can review and edit it before it reaches customers.

"Companies are trying to move more contacts to the Web in order to handle more customers and cut costs," Chuck Sykes points out. "But to answer more e-mails, they need more people, since automatic responses can't generate a human interaction with the customer, which they can't afford. You can get three times as many people for the same [cost] in India or the Philippines. Then, once e-mail numbers are up, you can start to use additional technology and expand services, channels and levels of interaction to save money."

Contracting customer care to offshore locations poses significant challenges. With outsourcers in countries thousands of miles away from the US, visiting support centers to check on operations is more costly and time-consuming than going to facilities in the US or Canada.

"If it is difficult to get clients to go to Omaha, how much more difficult will it be to have them go to Bangalore?" asks Oetting's Gantman.

Checking references is easier with domestic service bureaus than with off-shore firms, many of which have limited experience with outsourcing. Also, concerns about compliance with contracts, and whether you can entrust your customers and their data to third parties, are greater when partnering with overseas outsourcers. Many offshore service bureaus are headquartered in developing countries where the laws, in practice or in their breach, favor the home team.

"Do your due diligence, like you do when seeking a service bureau, but even more so when shopping for one internationally," advises Paul Kowal, president of consulting firm Kowal Associates (Boston, MA). "Check out their operations. Get the names of other clients and interview them. Test the call centers before committing. A service bureau support center that is 17 hours away poses more of a challenge to get to and troubleshoot than a center one hour away."

How To Outsource

Should you decide to outsource, you must check out prospective companies carefully. Decide what functions you want them to provide. Many outsourcers that offer support to clients' customers can also help out with inbound sales, and cross-selling and up-selling. Verify whether they employ the same reps for different functions; support reps are not necessarily good at sales and vice versa.

Pay special attention to reps' qualifications and training because they will be communicating highly specific information to customers.

Get to know how outsourcers bill. Some outsourcers that provide support bill by the minute or by the hour. Others pay reps based on incentives for meeting certain goals, like handling a certain number of calls or achieving sufficiently high customer satisfaction scores.

If your customers are consumers, expect to pay hourly rates of $30 for out-sourced level-one support. Outsourced support to businesses, which Gantman says involves greater complexity and urgency, increases to between $35 and $40 per hour. Most outsourced level-two support costs $40 to $45 per hour.

Contracts for support, like other types of outsourcing, usually run between one and three years; the longer the better.

"The benefit of entering into longer-term contracts, with reviews and renewals each year, is that to the company it gives lower rates and the benefit of longer-term employees," explains Gantman. "To the outsourcer it provides stability."

If you outsource support to accommodate demand, determine the minimum number of people you need on staff to maintain a core group, including experienced supervisors who can take seasonal staff up a learning curve quickly.

When outsourcing support, do the same kinds of due diligence as you would when outsourcing customer care and sales. Draft and send out requests for proposals. Get references. Do site visits. Test outsourcers on small projects. These can involve evening shifts, weekend shifts or situations when your center receives more requests for support than it can handle. Ideally, you want to try out outsourcers on established products where you have a track record to judge against.

Check to see if the service bureaus have any certification, but don't rely on that as your only criteria.

"Using support center certifications as a way of prescreening potential suppliers makes a great deal of sense," says Gantman. "However, certification should not take the place of careful and thorough selection processes."

Consider more than one outsourcer. If you prefer, several outsourcing firms can be brought on board to support your customers. Or you can split customer service and support among multiple outsourcers, as long as they are capable of providing support. This way you keep outsourcers hungry and honest. But, still keep some of your support in-house so that your customers are never stuck.

Outsourcing Case Study

Outsourcing can be an effective adjunct to a company's support operation, as the example of computer manufacturer micronpc.com shows. The company outsources its first-level support calls to ClientLogic (Nashville, TN), which assist customers by phone from three sites in Albuquerque, NM; Las Vegas, NV; and Toronto, ON, Canada. Reps with proven writing skills respond to customers' e-mail messages.

At the Toronto center, reps take calls from Canadian customers and those who speak languages other than English, plus overflow calls from the US. Otherwise, the three sites collectively run as one center, and operate 24x7.

The outsourcer also answers calls at night from the computer manufacturer's business and government customers. In addition, ClientLogic dispatches technicians for micronpc.com's customers who are entitled to on-site service, and, if necessary, arranges for delivery of parts.

The number of reps at ClientLogic who assist the computer manufacturer's customers varies according to demand. micronpc.com usually receives fewer calls during the summer and more calls during the winter. The company also experiences a spike in calls after it introduces new products or releases upgrades to existing products.

"During our slow time of year, ClientLogic moves our dedicated agents, who we would have kept idle in-house, to other projects and then back to us again," says Witherel. "We've had people who left our project [during slow periods] and then came back."

ClientLogic has a two-tier support team. If reps cannot answer customers' questions, they ask for assistance from product support specialists. If product support specialists cannot help, reps direct customers to supervisors. Only as a last resort does ClientLogic escalate calls to second-level reps at micronpc.com.

A very low escalation rate is but one indication of how efficiently ClientLogic supports micronpc.com's customers. Dave Witherel, manager of technical support for Micron Government Computer Systems, believes the two companies work together well because they think alike.

"Its [hiring] profile is so similar to ours that we only get quality people out of it," he says. "It has a very low turnover rate, less than 5%, which is excellent. When it hires for our program, it advertises us as one of its clients. And we're proud to say that ClientLogic is a partner of ours."

ClientLogic was not the first outsourcer micronpc.com tried. It contracted out portions of its technical support to other service bureaus before awarding ClientLogic approximately 20% of its consumer calls in March 1999.

The outsourcer began taking calls for micronpc.com in May 1999 at its Las Vegas call center, and largely due to growth, shifted another 20% of its calls to Albuquerque in August 1999. (ClientLogic opened its Toronto center in October 1999.)

By February 2000, micronpc.com had discontinued its collaboration with other outsourcers. The manufacturer now directs all its consumer customers to ClientLogic.

"At one time we thought we shouldn't keep all our eggs in one basket," says Witherel. "But with ClientLogic, we could have the benefit [of multiple outsourcers] while staying with one outsourcer because we could compare its performance across the different centers."

The partnership is paying off for both companies. The outsourcer has been able to handle an increasing number of calls and resolve a greater percentage of requests for support than micronpc.com could on its own — all while keeping costs in line.

Witherel adds that the outsourcer puts much more effort into monitoring agents than micronpc.com asks for, and evaluates more calls than other service bureaus he has worked with.

Monitoring is only as good as the quality of the reps, which in turn depends greatly on the training they receive. The reps at ClientLogic who support the computer manufacturer's customers undergo a 17-day program that includes training from micronpc.com. The reps use equipment from the manufacturer, and they rely on the same diagnostic, trouble ticketing and knowledge base software as the second-level support reps at micronpc.com. The outsourcer also furnishes each of its centers with micronpc.com product labs.

"We have always felt that in order to give our customers via ClientLogic the best service possible, we always need to provide reps with the best training possible," says Witherel.

Insourcing

When you insource, you contract out your support center operations to a company, usually a staffing agency, that recruits, screens, trains and supervises reps on your premises. Although insourced reps typically work for less money and fewer benefits than reps you hire, they can come and go with greater ease; and you do have the option of hiring insourced reps to work for your company after a certain period of time.

Insourcing is very common for functions besides support. For example, most companies insource their mailroom, janitorial and security work. In his youth, co-author Brendan Read worked as an insourced security guard at many offices, factories and construction sites throughout Canada, as well as the 1978 Commonwealth Games in Edmonton.

As with outsourcing, insourcing offers you flexibility. You tell the insourcer how many people you need, when you need them and what skills they need to have.

Insourcers are great resources for setting up new support centers. They're familiar with the labor pools in different communities. They know how to get people to join and stay with your support center.

Companies often insource support staff to get a new center up and running. After several months, they shift over the staffing and management to in-house personnel.

With reps from insourcers, you have greater control than you do with outsourcers. Reps are using your systems, so you avoid having to integrate products that your outsourcers are using. You also avoid potential technology compatibility problems that sometimes occur with outsourcers because the insourced reps are using your gear.

Perhaps the most important advantage of insourcing is that if there is a problem with any insourced rep, all you need to do is to alert the insourcer and the rep no longer has to work at your center.

The downsides of insourcing are that you still have to build and supply the facilities, which outsourcers ordinarily take care of. You're also limited to the labor force near your support center, whereas outsourcers can select locations where the labor force is qualified and affordable.

You also entrust the hiring and screening to the insourcing agency. The reps are the agencies' employees, not yours, which means you have less control over them than if they were your company's employees. Insourcers tend to attract individuals in the midst of career and life changes. Therefore turnover among insourced reps is usually greater than it is with in-house reps.

As with outsourcers, make sure to research prospective insourcers with care. Get references, and test out insourcers on short-term projects first to see how they perform.

Six

CHAPTER SIX

LOCATING AND BUILDING SUPPORT CENTERS

Locating and Building Support Centers

CHAPTER Six

For many companies, the volume of requests for support warrants only a few reps. But when your demand for support requires 20 or more reps at any one time, then you have to maintain a full-fledge *support center*.

Support centers represent the elite among corporate call centers. They require reps with specialized training to assist customers with anything from power windows to Windows operating systems.

Labor comprises about 70% of a support center's operating costs, a higher percentage than you're likely to find at other types of call centers, where labor typically accounts for 60% of costs. You have to pay a premium to attract reps that can communicate knowledgeably and effectively with customers.

Consider, for example, the level-one reps whom your customers initially reach, and who, according to experts, handle 75% of support inquiries. Expect to pay 10% to 15% more for level-one support reps than for reps in call centers that do not provide support.

You have to take special care in selecting sites where you employ support reps, whether you run a service bureau or your company's in-house support center. As is true of call centers in general, wages depend on the labor market. They are typically higher in major American metropolitan areas, less in smaller cities and lower still in other countries, like Canada. Outsourcers typically pay less than in-house support centers.

King White, a vice president with the Dallas, TX-based Trammell Crow Call Center Site Selection Group, observes that support centers pay between $13 and $15 per hour for level-one reps in major US cities. By comparison, many tech-

Customer Contact Center Relocation Savings Analysis

	Sunnyvale, CA	Alternative Location[4]
ESTIMATED LABOR COSTS:[1]		
Total CSRs	100	100
Annual Salary	$55,000	$40,000
Average Wage Rate	$26.44	$19.23
TOTAL ANNUAL LABOR COST	$5,500,000.00	$4,000,000.00
LABOR COST VARIANCE	37.5%	
EMPLOYEE ATTRITION COSTS[1]		
Estimated Annual Turnover	5.00%	10.00%
Average Training Costs Per CSR	$9,000.00	$6,000.00
TOTAL ANNUAL ATTRITION COSTS	$337,500.00	$450,000.00
ATTRITION COST VARIANCE	-25.0%	
ESTIMATED OCCUPANCY COSTS:[2]		
Estimated Square Footage	20,000	20,000
Annual Full Service Rental Rate	$72.00	$22.00
Annual Rental Costs	$1,440,000.00	$440,000.00
Additional Cost (i.e. parking etc.)	not disclosed	n/a
TOTAL ANNUAL OCCUPANCY EXPENSE	$1,440,000.00	$440,000.00
TOTAL OCCUPANCY COST VARIANCE	227.3%	
TOTAL ANNUAL EXPENSE:	$7,277,500	$4,890,000
ANNUAL EXPENSE PER EMPLOYEE:	$72,775	$48,900
TOTAL ANNUAL SAVINGS:	$2,387,500	
ANNUAL SAVINGS PER EMPLOYEE:	$15,917	
TOTAL FIVE YEAR EXPENSE:	$36,387,500	$24,450,000
TOTAL FIVE YEAR SAVINGS:	$11,937,500	
TOTAL COST VARIANCE:	32.8%	
POTENTIAL ADDITIONAL EXPENSES:[3]		
New Switch	n/a	$500,000.00
UPS ($250,000 per 250 workstations)	n/a	$250,000.00
1 Meg KVA Back-Up Generator	n/a	$120,000.00
Voice/Data Cabling ($300 per workstation)	n/a	$30,000.00
Desktop Computers ($1,250 per workstation)	n/a	$24,038.46
Furniture ($2,500 per workstation)	n/a	$250,000.00
Additional Furniture (cafeteria, training, lobby)	n/a	$30,000.00
TOTAL POTENTIAL ADDITIONAL EXPENSES:	$0.00	$1,204,038.46

This analysis does not include savings achieved from economic incentives negotiation.
Many alternative locations will provide new employee training assistance,
tax abatements and cash incentives as incentives for relocation.

[1] Estimated labor rates and turnover rates are based on Site Selection Consulting Group experience
and research in a Tier I market with heavy concentration of IT workforce.

[2] Occupancy costs are based on rental rates in the Sunnyvale area vs a Tier I market described above.

[3] Some potential additional expenses may be depreciated over a 5-10 year period if incurred.

[4] The alternative location used in this analysis is a Tier I market with a heavy concentration of an IT workforce
and recent college graduates.

Trammell Crow Company
Site Selection Consulting Group

High-end support centers are not inexpensive. But, as this analysis by Trammell Crow for a leading client illustrates, you can save millions in support costs depending on where you locate your center.

nical support outsourcers pay $8 per hour in the US and $6 per hour in Canada.

Given the higher labor costs, support centers can be expensive to build — as much as 10% more than customer service call centers. The cost per seat runs about $35,000. Real estate expenses range from $300,000 to more than $1 million annually, depending on location.

For more general background on how to design and select sites for call centers, consult <u>Designing the Best Call Center for Your Business</u> also published by CMP Media.

Planning

In sizing your support center, as with other types of call centers, you need to account for how many calls and how much on-line correspondence reps receive. But don't look at quantity alone. In terms of volume, the bulk of a company's dealings with customers involve sales and service rather than support. Compared to other requests from customers, though, those for support tend to take up more time, and entail different staffing requirements.

Efficiency is relative. With the help of providers of benchmarking services (see chapter 9), you can identify businesses that are similar to yours, including competitors within your industry. Then you can compare key metrics, like resolution times, for these businesses' support centers with those of your center.

Make sure your pricing reflects customers' demand for and satisfaction with your support. If you offer live help to customers who purchase commodities like computers or clocks, surveys can help you find out if customers are willing to pay a little extra, and are more likely to buy from your company again, based solely on how good your support is.

If it's determined that it's feasible to charge a per minute fee for support and you institute a fee-based system, you'll find your customers more eager to resolve their problems quickly, enabling your support center to handle more calls per rep than the average support center. Just be wary. You don't want your pricing policy to alienate customers. (See chapter 3 for more on fee-based support.)

Strategies

When setting up a new support center, you have to develop a strategy to ensure optimal use of your resources.

Automation is one of the tools that can help you execute on this strategy.

Look carefully at how you classify and handle requests for support. For example, if you'd prefer to reserve reps to deal with the customers' most challenging questions, you can allow customers to reach reps only after they search for answers themselves with the help of an interactive voice response (IVR) system or an on-line knowledge base.

❖ **NOTE:** Chapters 7 and 10 provide a thorough discussion of knowledge management, knowledge bases and call center technology in general.

Forecasting is another tool that enables you to plan according to your strategy. But, forecasting requires more than figuring out how many reps with certain skills you need at any given time. If you have more than one support center, forecasting tools can help you effectively distribute your staff. And it presents you with the most efficient allocations among in-house, outsourced and at-home reps.

If there are seasonal variations in call and contact demand, like Christmas (if you make consumer electronics) and Father's Day (if you make electric razors), you can reasonably predict them and use forecasting technology to plan and staff your customer support operations (on-premise and teleworkers) and negotiate for in your outsourcing contracts.

But you will also need to plan for the unexpected — higher than expected sales, bug fixes and product recalls — based on past experiences. You will then need to have a strategy, like outfitting training stations, and arranging for these eventualities with staffing or temp agencies (more about them in chapter 8) or outsourcers (see chapter 5).

There are multiple advantages of maintaining more than one site. With centers in different time zones, you allow customers to receive support at times that are convenient for them without having to struggle to hire reps for late or early shifts. You can follow the sun — closing a support center in one time zone, like in Portland, ME at 11 pm and transferring them to a support center in another time zone, like in Portland, OR where it is only 8 pm. When your Oregon center is closed, say at 4 am, your support center in Maine can be open to help the early birds catch those worms in time. Like fixing stock market software before the Wall Street bell rings.

You're also more likely to find reps who are collectively willing to assist customers from, say, 8 am to 9 pm Eastern time, if you have a center in California and New York than if you have only one site in one time zone. With a mix of teleworkers, outsourcing and multiple site, you also can better respond to any kind of disruptive event (floods, storms, earthquakes, power outages) and recover quicker from a disaster.

Outsourcing and teleworking are also effective strategies for after-hours support or ad hoc support during times of peak demand. Allowing reps to work from their homes enables you to find staff during tough-to-fill evening and weekend shifts. Both methods enable you to cut down on the number of requirements for physical space for your support center.

In many companies, there is a strong synergy between support, engineering and development that works to everyone's advantage. So, you might want to take a look at co-locating your support with these departments. Co-locating is common practice in small companies or departments, but it also occurs in large firms, like Mitel and Xerox.

Design Requirements

Customer support facilities need higher-quality furnishings than a customer service or sales center, according to Roger Kingsland, principal with KSBA Architects (Pittsburgh, PA). For instance, since support reps spend more time communicating with customers than other reps they need more privacy and soundproofing.

Kingsland says that because support reps tend to earn higher pay, they expect more from their surroundings. Interiors must especially reflect the preferences of the labor force, which in many, but not all, support centers is comprised of young people between the ages of 20 and 25. The center must use contemporary colors, lighting fixtures and design elements.

Support reps also need additional room. Compared to reps in most call centers, where between 65 sf and 80 sf per rep is sufficient, support reps need between 80 sf and 100 sf to accommodate bookshelves, cabinets and manuals. If you provide support for tangible items, like the new DVD home theatre systems, hi-tech refrigerators or printers, you may need to set aside space for reps to have these items on hand when customers have questions about specific products.

The lighting needs of support centers are the same as for call centers: about 30 foot-candles, which is at least half the number of foot-candles of lighting in typical offices where the range is between 50 and 80 foot-candles.

Support reps spend a lot of time going over documents such as manuals. To enable reps to see the fine print, provided you haven't already replicated these documents within your intranet, you may need to supplement indirect lighting with lamps and other types of task lighting.

How you lay out your support center depends in part on the support level. For instance, level-one reps usually work alone with occasional assistance from

Co-Location

If your company supports high-tech products, take a close look at co-locating your support center with other business functions, such as engineering and product development.

With co-location, engineers and developers get to hear firsthand customers' experiences with the company's products and services and, thus, can respond quickly and accordingly with fixes and new releases. In turn, the support centers are able to get on top of problems swiftly and accurately if customers identify flaws with a product.

Co-location also offers another important benefit for support desks: It gives ambitious reps career paths within the same company. Offering advancement in a co-located facility is a superb way to attract top quality candidates.

Companies that look to locate more than one function have bigger clout with landlords and developers. They can strike better deals if they lease the entire building than if they rent one or two floors. Co-locating eliminates parking hassles with other tenants and provides better security.

When examining co-location, verify that the site you're considering attracts sufficient quality workers at the right price for all functions. Also make sure the site is near other firms and professionals.

supervisors. You can accommodate these reps in higher-density rows of cubicles.

You have several choices for how you set up your reps' workstations. Several furniture manufacturers offer core or pod layouts, where reps sit facing a central core. Alternatives to boxy cubicles include curving and zig-zagging workstations.

Jack Warden, project director and architectural team leader of WorkPlace USA (Dallas, TX), says his customers are staying away these nontraditional designs because they cost 15% more than cubicles and take up 15% to 20% more space.

"End users have backed off the fancy design," says Warden. "They have not proved more effective at retaining and attracting agents. You can make cube rows acceptable in regional or corporate headquarters by using more muted col-

For example, Sunnyvale, CA, in Silicon Valley, and Waltham, MA, on the fabled Route 128 'electronics belt' are great locations for co-locating a support center with engineering, product development, administration and management.

Yet, certain areas, despite their association with high-tech, are prohibitively expensive. Trammell Crow reports that some Silicon Valley support centers pay annual salaries of more than $55,000 to reps, compared with salaries of $30,000 or $40,000 elsewhere (see chart on this chapter). Real estate costs in Silicon Valley are extremely high, at $72 per square foot, compared with $18 to $23 per square foot (sf) in other communities.

One way around this issue is to co-locate your support center where you have regional offices. For this to be a viable strategy, you need to examine which labor markets fit your budget and match them with where you need offices. If you find a fit, you can achieve savings in both real estate and labor. And chances are the regional offices employ engineers and developers so that you can achieve the synergies you desire.

You can save between 10% and 30% by co-locating, say, a support center and a warehouse, compared with keeping various operations separate. Staff can share cafeterias, on-site power and parking.

The actual amount saved depends on factors like the sizes and functions of the business units you locate together. Other factors to keep in mind are whether the facility is a renovation, whether the site chosen for dual duty is an office or factory, and what on-site options, like parking and utilities, come with the location.

You can make reps' cubicles efficient and stylish with designs like Interior Concepts' (Spring Lake, MI) Swurv line.

ors, like grays and tans; and by using higher-quality fabrics. It comes back to density, to do more with less."

Design needs are also different in centers that need a more collaborative environment, says Kingsland. If your center stresses teamwork among reps, alternatives to cubicles include partition-less workstation rows, where reps can move from workstation to workstation, or a central conference table between rows.

Ergonomics

Repetitive motions like typing, poor posture, and not enough time for breaks can exacerbate conditions like carpal tunnel syndrome and tendonitis. To prevent most carpal tunnel injuries, your center only has to spend about $350 more per workstation, for items such as for adjustable chairs, keyboards and workstations and wrist pads.

Reps should receive training on how to use these items.

"You don't put agents on a computer or a phone and tell them to use it without training," KSBA's Kingsland points out. "So you shouldn't provide reps with adjustable chairs, workstations and keyboard rests without showing them

Split-PCs for Your Support Center

When planning to fit out a new support center or upgrade your existing center, instead of plunking down desktop PCs, consider installing split-PCs. With split-PCs, the central processing units (CPUs) sit mostly or partially in computer rooms, connected to keyboards and monitors on desktops. If one PC fails, it doesn't take the other terminals and workstations with it. Your IT staff can pull out the old one and drop in a new one.

These features make split-PCs very scalable compared to network PCs, 'thin clients' and dumb terminals that feed into one or a few servers. You add and remove units as needed. According to the experts — split-PCs are more reliable than the typical desktop PCs.

Split-PCs take two forms: specially designed and off-the-shelf. Clear Cube (Austin, TX) is one manufacturer of specially-designed split-PCs. One Clear Cube rack, for example, holds 96 CPU blades. A consultant or systems integrator can install and disassemble off-the-shelf units.

Either way the benefits are the same. They include less wiring and installation costs and time. A split-PC requires only one category-5 cable to each workstation compared with three or four cables to desktop PCs, plus phone cabling.

Split-PCs may also reduce air conditioning costs by limiting heat from cabling to every workstation; and by handling heat through centralized computer room AC systems.

Split-PCs enable greater control and less upgrade time and costs. IT staff can upgrade computers centrally, instead of modifying every workstation's CPU. Also reps cannot load extraneous software and you minimize CPU theft and vandalism by disgruntled employees.

Kingsland Scott Bauer Associates (KSBA; Pittsburgh, PA), which designs call centers, would like to see a complete cost benefit of split-PCs. "Think of split-PCs like long monitor and keyboard wires from a CPU that could be anywhere in your office," says KSBA principal Roger Kingsland. "The CPU doesn't have to be where you are to enable you to work on it."

how to use these features."

Making your support center ergonomically sound pays off because it avoids injuries that would otherwise increase sick time, lower productivity and raise workers' compensation claims.

"According to OSHA, the average claim [from injuries like carpal tunnel] is $35,000," says Christine Jacobs, marketing director with furniture maker Interior Concepts. "For the same amount, a call center can outfit a 100-seat facility with workstations that will help prevent that from happening."

If your company provides technical assistance, support reps may need more equipment, which means more wiring. To ensure your center can set up and relocate workstations quickly and easily, a flexible wiring system is best.

One example is raised access flooring (RAF), where the cabling is under the floor rather than in the ceiling. It's the opposite of conventional cabling for cubicles, which feeds to workstations through power poles that connect up to wires located above ceiling tiles.

The principal downside to RAF is its higher cost. But according to architects like Roger Kingsland, you can cut capital expenses and improve environmental quality within your center if you integrate electrical wiring with cabling for heating, ventilation and air conditioning.

Another method of flexible cabling uses panels, such as Herman Miller's (Zeeland, MI) *Resolve* which extends cubicle partitions to reach wiring within your ceiling. But, Kingsland warns, there are downsides. Furniture panels add another $60 to $150 to the cost of each workstation and a change to one wiring panel affects the other panels. That's not the case with modular, in-floor cabling systems.

In terms of monitors, flat panels are worth considering because they take up less space than conventional computer monitors.

Kingsland conducted research on flat panel monitors for IBM. He discovered that although flat-panel monitors each cost between $300 and $600 more than conventional monitors, they reduce real estate expenses, given their lesser depth. For a rep's cubicle, the extra space adds up to between six and ten sf per workstation. In a 600-seat center, the additional space can save you half a million dollars.

Flat panel monitors also consume less energy, and since they don't flicker as much as other types of monitors, they reduce eyestrain as well.

The amenities your support center needs depends on the labor force that works best in them, and the community you're locating into. If your workforce is young and likes to cycle, install bike racks outside or inside your foyer, and have large washrooms for them to change in and out of their cycling gear.

Consult with your reps. Test amenities out in your existing support centers before investing in them in your new facility. While reps may say they would enjoy having a gym in a new center, that doesn't guarantee they would use it.

Don't Cheapen Design

Being pennywise makes sense in design. But don't risk the temptation to be pound-foolish.

Bob Engel and Gere Picasso, principals with Engel Picasso Associates (Albuquerque, NM), point out that about 90% of the operating and capital costs in call and support centers are associated with personnel, while only 10% are spent on facilities. By disproportionately focusing on where you can save in that 10%, you'll pay much more in the long run. Cheapening your design, like putting in nonergonomic workstations, may prove to be costly — more sick time, lost productivity and greater turnover.

There are some very innovative designs for support centers. The Sustainable Technology Business Center, which Kingsland Scott Bauer Associates (Pittsburgh, PA) developed for Dallas-based Hunt Power's CentraTek subsidiary, features plenty of natural light through clerestory windows, skylights and side windows. This light reaches nearly two thirds of the floor area. Heating, ventilating and air conditioning units sit on overhead equipment rails for easy addition and maintenance. Ducts located under the floor feed in fresh air and remove exhaust.

Engel and Picasso cite a hypothetical 500-rep support center where installing small, substandard nonergonomic workstations can easily result in a 10% increase in turnover, which they say is an extremely low estimate. This hypothetical center must now recruit and hire at least 50 new reps a year above its norm. That typically translates into another $30,000 in recruiting, training and starting wages and benefits per rep, or $1.5 million per year.

"We don't think that any savings in furnishings or space rents could ever counterbalance this $1.5 million in additional personnel costs," says Picasso. "If the additional turnover is 20% to 30% created by facility 'savings,' then each year $3 million to $4 million is thrown away, more than your total annual bill for facilities, furniture, technology and utilities."

> ❖ **NOTE:** Many design features and considerations for support desks are applicable for other functions, like administrative, back office, engineering and product development. For example, flat panel monitors, raised access flooring, ergonomic and flexible furniture can and should be used for all departments.

Site Selection

Labor is the most important factor to look for where you seek to locate a call center. But you must qualify that with quality labor for support.

It's possible to teach people to provide support effectively. For example, support outsourcer Stream International has hired ex-coal miners at its call center in Glace Bay, Nova Scotia, Canada, and has apprenticeship programs at another Canadian facility in Chilliwack, British Columbia. It has reached out successfully to the Sto:Lo First Nations in Chilliwack; the aboriginal band recruits and trains applicants.

The more sophisticated the support, the higher the qualifications and the more limited the labor pool. If you need reps who have certification on a branded Unix platform or who knows Linux, don't expect to find them in a farming community where Unix machines are scarcer than hen's teeth, or where people would pronounce or spell "Linux" the same as the name of a famed Charles Schultz cartoon character.

If your center provides technical support, don't be deluded into thinking that you're going to find all the reps you need in Silicon Valley in California or in Silicon Alley in New York. The labor market in these locations is more costly than in other locations, and you're more likely to attract applicants for IT jobs than for positions in your center.

Consider the applicant's point of view: Why should he or she work in support instead of earning more by developing the software or equipment you're supporting? Moreover, consultants warn that by locating in tech-rich communities, other companies may use your firm as a training ground for support reps and developers.

In all support levels, the communities you pick must have good colleges that offer training, whether it's for workers who had been in another field all their lives or for placing young people into careers. College students comprise an excellent pool of part-time reps, as well as reps for summers and holidays. Ideally, you should locate your support center where it is accessible to schools by public transportation, by car or by bike.

If your workforce is primarily young people, take a close look at locations at or

Locating Your Support Center

Putting your support center in the right labor market saves money. King White, vice president of call center services with Trammell Crow (Dallas, TX), uses the example of a 300-seat center that employs the equivalent of 300 full-time level-one reps and occupies 40,000 square feet at $23 per square foot. The annual turnover at the center is 55%, and reps earn $15 per hour.

Assuming a 40-hour work week, the annual costs are $9.36 million for labor and $920,000 for the lease on the property. To cut costs, the company finds another community where it can pay $12 per hour, or $24,960 per year. The wage bill drops to $7.49 million. That saves about $1.9 million in payroll and pays for your entire facility for two years.

With lower wages comes lower turnover costs, which leads to further cost savings. Let's say the average turnover rate is 20% in the new location, which means that the equivalent of 60 full-time reps leave each year. The costs of turnover represent about 50% of a rep's annual wages, in terms of recruiting, screening, training, lost productivity and other expenses. With 20% annual turnover, the cost is $249,600; with 55% turnover, the cost is $858,000, an increase of 244%.

In cities with lower wages, property costs are often lower, too. For example, if your lease within a city drops from $23 to $18 square feet, you save $290,000 per year in rent. But the lesson is clear: Given a choice of lower rent or lower labor costs, provided the quality of the workforce isn't lower, you'll save more in the long term in cities with lower wages.

near downtown locations, where there is activity and nightlife that appeals to them. In these areas, reps also can save on parking costs, as many cities have new and rebuilt public transportation systems or bikeways.

Also, take a look at your support center's size. You may find you have more choices of sites if you take, say, a 300-seat center and split it two or three ways.

Property

The big issue when picking property is parking and access. Support and call centers typically require large amounts of parking, with ratios of five or six spaces per 1,000 sf because their occupancy density is greater than other office uses and because they often have shifts that will need more spaces to accommodate cars pulling in and out.

The parking space a support center needs also depends on the community and

Comparing American and Canadian Locations										
	San Diego, CA	Dallas, TX	Charlotte, NC	Tampa, FL	Orlando, FL	Toronto, ON	Montreal, QUE	Ottawa, ON	Edmonton, AB	Saint John, NB
Nonexempt Labor										
Weighted Average Monthly Earnings	$2,375	$2,360	$2,316	$2,213	$2,134	$1,956	$1,828	$1,776	$1,687	$1,574
Annual Base Payroll Costs	$8,550,000	$8,496,000	$8,337,600	$7,966,800	$7,682,400	$7,041,600	$6,580,800	$6,393,600	$6,073,200	$5,666,400
Fringe Benefits	$2,821,500	$2,803,680	$2,751,408	$2,629,044	$2,535,192	$1,337,904	$1,250,352	$1,214,784	$1,153,908	$1,076,616
Total Annual Labor Costs	$11,371,500	$11,299,680	$11,089,008	$10,595,844	$10,217,592	$8,379,504	$7,831,152	$7,608,584	$7,227,108	$6,743,016
Electric Power Costs	$61,639	$41,832	$38,393	$41,556	$42,073	$33,417	$25,909	$26,398	$24,320	$32,631
Office Rent Costs	$718,148	$644,988	$572,536	$485,688	$589,764	$348,100	$389,400	$454,300	$247,800	$312,700
Equipment Amortization Costs	$1,132,800	$1,132,800	$1,132,800	$1,132,800	$1,132,800	$1,132,800	$1,132,800	$1,132,800	$1,132,800	$1,132,800
Heating and Air Conditioning	$23,140	$41,116	$31,614	$42,178	$40,637	$43,346	$38,559	$39,621	$38,739	$47,008
Telecommunications Costs	$1,209,437	$1,203,246	$1,208,853	$1,269,427	$1,269,427	$1,249,974	$1,249,974	$1,249,974	$1,249,974	$1,249,974
Total Annual Geographically-Variable Operating Costs	$14,516,664	$14,363,662	$14,073,204	$13,567,493	$13,292,293	$11,187,141	$10,667,794	$10,511,677	$9,920,741	$9,518,129

Notes: Includes all major geographically-variable operating costs. Reflects a 36,000 sq. ft. call center in the financial services sector employing 300 nonexempt workers and having a monthly call volume (inbound) of 2.7 million minutes. All figures are in U.S. dollars at an exchange rate of $1.50

Source: The Boyd Company, Inc., Princeton, NJ

There is wide variation in capital and operating costs for support centers by city, as this chart from The Boyd Company (Princeton, NJ) points out. And it may be less expensive to locate your support centers in Canada. The simulation is for a 36,000-square-foot corporate support center with 300 reps and a monthly inbound call volume of 2.7 million minutes.

the workforce. If the reps your support center wishes to attract are college students, many of them may commute on bicycles, so you may need more cycle racks than parking spaces. If the workforce you seek is in a city with a trendy downtown, and reps are willing to take mass transit, then you can lower or in some cases do away with parking requirements.

SafeHarbor's Support Center Glows

Support outsourcer SafeHarbor (Satsop, WA) likes to be on the cutting edge of customer support.

This can be seen, literally — SafeHarbor has its headquarters and support center in a never-activated former nuclear power station in central Washington State, now the Satsop Development Park. Many call centers and support centers have located in former factories, but to our knowledge, this is the first time anyone in the US has built one on a nuclear site.

The outsourcer picked the location, which has 208 workstations in 44,000 square feet, because the company sought to contribute to the economic development of the region. SafeHarbor became the park's first tenant, in 1998.

The outsourcer cites three reasons for its choice of location. First, the site offers tremendous fiber optic connectivity and a substantial power infrastructure. Second, the region has a stable supply of motivated workers. Third, SafeHarbor received significant financial and educational incentives from federal, state, and local governments.

SafeHarbor contracted with Harbor Architects (Aberdeen, WA) to renovate the structure, including a health club for employees. The layout of SafeHarbor's workstations facilitates collaboration between reps who communicate directly with customers and "knowledge engineers" who maintain an on-line knowledge base for clients.

Consider your power sources, and don't become too dependent on your local utility. Daniel Frasca, vice president of The Alter Group (Skokie, IL), reports that more companies are requesting on-site power generators for their call centers, with the commercial power grid, rather than the generator, as the backup.

You should also take a hard look at security. Since the terror attacks, firms are tightening building security. They are also more conscious about providing lighting and removing shadows at night to enable workers to walk safely from the building doors to their car doors.

If you pursue outsourcing and teleworking, the reps who work from their homes or for outsourcers can assist customers during evenings and weekends. This gives you the option of closing down and locking up your center at these times, thereby reducing security risks.

When selecting property, you may wish to look at resale value, known as "residuals." Retail and factory conversions are less likely than conventional offices to increase residuals.

According to The Alter Group, companies are more willing to invest in more high-quality exterior design, such as all-glass sides instead of standard finished concrete and glass. Companies gladly pay for ribbon windows or windows punched through concrete if they net them better residuals.

All-glass exteriors are 10% to 15% more costly than these other combinations; tinting glass reduces the extra heating costs. The payoff comes in a higher-profile corporate image to attract employees, clients and greater residuals. The price of glass is dropping, although it's still a bit more expensive than concrete construction and finish.

"Our clients are beginning to understand that if they have a more attractive building, which glass achieves, that they could sublease at a higher rental rate," explains senior vice president Kurt Rosene of The Alter Group. "And [they can] get it subleased quicker than a standard concrete and glass building in the same location."

Choosing a Community

Because locating and operating support centers cost a lot of money, you must examine the communities that have the best mix of people and property.

Don't limit your choices based on preconceived notions. Call centers, such as those for credit card giant MBNA, have chosen sites in once-blighted areas like Newark, NJ and Belfast, ME. Yet many companies, often out of fear, stay away from inner cities or areas of high unemployment. That's despite incentives from these cities to locate there, as well as large numbers of people who are smart and eager to work.

For sound labor and economic reasons, many support centers that support American customers are outside the US. Countries like Barbados, Canada, Costa Rica, India, Panama, The Philippines, and provinces like Northern Ireland are vying for customer support centers.

Many countries' cities have higher unemployment rates and greater availability than their US counterparts, with labor costs that run from 30% to 75% less than in the US. Consequently, turnover tends to be lower, saving staffing and training expenses while delivering better performance as agents build up experience and knowledge.

Many countries also have large pools of well-educated people willing to work in support centers. One reason is that these countries have better education systems than in the US. Another reason is that some countries offer few comparable employment opportunities at the wages call centers pay for people with college degrees. That means you can hire someone with a doctorate for the same, if not less, money than someone with a high school diploma in the US.

For many companies, the labor availability and quality alone are enough to prompt them to locate call centers in other nations. Support centers that are set up in other countries often attract and keep applicants with college degrees who can read and write at a level sufficient for on-line communication.

"We found that the labor costs in Canada are not necessarily lower than similar areas in the US, especially when you add in Canadian taxes and telecom costs," explains Kathleen Nordgren, a spokesperson with outsourcer Stream International. "But the real driver that led us to open there and to consider other locations within Canada is the highly-skilled and available workforce. With Canada being on the North American dialing plan, call routing is not much different than that in the US."

As more people emigrate to the US and retain their cultures and languages, Americans are slowly becoming used to speaking with others with different accents. Customers could conceivably be speaking with support reps in Kingston in Jamaica, Ontario or New York.

But locating abroad comes at a price. The logistics and time required to maintain centers outside of the US is greater than within US borders. There are complicated and unfamiliar laws, practices, regulations and agencies to cope with, plus cultural and language differences. Some countries' bureaucracies are notoriously ensnarled in red tape, whereas others have more stringent labor laws or stronger unions than in the US. Also, travel to these locales can be time-consuming.

Many people in other countries speak English far faster and with accents too heavy for many Americans to understand. To get around this issue some firms split their support: answering e-mail in countries like India and answering calls in Canada.

"Culture and accent is not that great of an issue with basic first level support, because the calls and e-mails are relatively simple," points out Bill Rose, executive director with the Service and Support Professionals Association (SSPA; San Diego, CA). "But where you do get into problems with accents is with high-end first-level and second-level support, where the reps must engage in complex conversations with callers."

Setting up an Indian Support Center

India is a popular location for outsourced and, increasingly, in-house support centers. But there are special design issues companies must be aware of, as the experience of global outsourcer Convergys reveals.

The center has room for up to 1,900 seats across 200,000 square feet on six stories. Convergys initially began operations with 300 workstations and plans to eventually occupy the entire building.

During its search for a site, Convergys found the property, which had been slated for an office. But the prospective tenant had pulled out, leaving the foundations and design in place.

Dennis Ross, Convergys' manager of offshore operations, says that nabbing the property enabled Convergys to become operational several months faster than it would have if it had started building from scratch. Convergys was also better able to customize the building based on US standards.

The firm opened its first Indian call center for customer service and technical support, in Gurgaon, a New Delhi suburb, in October 2001. The center is located near the offices of other multinational firms including Coca-Cola, GE, Motorola and Nestle.

Convergys' plan for the Gurgaon center differed from the building and renovation of its US and Canadian centers. For example, Convergys had to put in bus loops and shelters. Although some Indian employees drive to work, most expect companies to transport them to work with shuttle buses and vans.

The outsourcer also installed motorcycle racks. Many support reps who get to work by themselves ride motorcycles, which are less expensive to buy and maintain than cars, and they run better on India's chaotic roads.

The outsourcer also put in a large full-service cafeteria, compared with the simple lunchroom and vending machines typical in the US or Canada.

"You don't have the culture nor the outlets for fast food or restaurants off-site in India as you do in the US or Canada," explains Ross.

Convergys' center in India, like many of its North American counterparts, has on-site diesel-driven electrical generators. Given that India's commercial power system is still not very reliable, the on-site generators are larger and have enough fuel to run the entire call center for up to 20 days, compared to typical US generators that run for seven days.

There are no switches and servers in the Indian facility. Instead, Convergys routes communication over privately-leased T-1 and E-1 circuits in the US. The configuration gives Convergys control of its end-to-end operations and processes, just as the company has in the US, Canada and Europe. It also cuts down on IT expenses at its Indian call center. Convergys also had its workstations custom-built, which is less expensive than factory-ordered workstations in India.

The call center uses high quality glass and marble because these materials cost less to purchase and install. And they help with cooling. (Marble is a "cool" material; it does not trap heat.) Tinted glass also reflects light, cutting heat.

Convergys had to put more space into the Indian call center than it provides in US and Canadian centers because, in addition to training reps on products, the company also shows reps how to understand American accents.

There are a few other interior differences. Indians tend to like more vibrant colors than Americans, including shades of blue, purple and red.

"One other difference we found when we built the center was in the building support pillars," notes Ross. "In the US, designers hide them with plasterboard and other coverings. But in India the agents preferred to have them uncovered and asked us to backlight [the support pillars] because they showed strength."

Reusing Support Centers

If you're looking for affordable space your company can easily move into, try reusing support centers other companies have already vacated. If the old support center has closed down fairly recently, you may still be able to attract staff whom the previous company laid off. The site may come with furniture and equipment, and may require only minimal construction, says King White, vice president of call center services with Trammell Crow (Dallas, TX).

But be warned: Moving into these centers carries risks. Many centers that are one or two years old are up-to-date; facilities older than that may not be. The furniture may be worn out or not up to modern standards. The equipment and wiring may be obsolete.

Even if the space is up-to-date, the design may be unsuitable. A site initially designed for a customer service operation may not accommodate space for reps who provide support.

By taking over a former call center, you may also inherit another center's problems with labor supply, facilities and technology, adds Roger Kingsland, principal with Kingsland Scott Bauer Associates (KSBA; Pittsburgh, PA).

"The former tenant short-listed that call center for a reason," he points out. "So before taking it over, find out why they left it."

Be cautious, too, about reusing furniture sometimes left behind by the previous tenant. Even if the furniture contains steel frames, key features like adjustable chair lifts, armrests and backrests do not. Nor does the upholstery. Bob Engel, a principal and founder of Engel Picasso Associates (Albuquerque, NM), warns that although used furniture costs less at first, bringing second-hand pieces up to modern standards may be more expensive than buying new furniture, especially if the old pieces are out of warranty.

Worse, the chairs and workstations may not be ergonomically sound, and could cause injuries to reps. Older workstations also may cost more to move around than newer models. Why? Units made before the late 1990s requires passages for electricians to pass through wiring. To move furniture, requires disassembly of the cabling network completely. Today's furniture features easily accessible trays where electricians can lay in wiring without taking apart the cable network or the furniture. This one feature can reduce cabling costs between $250 and $300 per workstation, compared with $500 to $600 with older furniture.

Lastly, the furniture may be outdated. If you have to add or replace your upholstery, you may not get the right fit of design, color and fabrics.

"If you don't plan to change around your call center and you don't care how it looks, then you can get away with older un-refurbished furniture," says Engel. "But if you do care about performance, appearance, flexibility, employee health and total cost, then you should buy new."

Language Issues

Special circumstances apply when you support customers from centers outside the US. Chances are that you will not find enough reps who speak your customers' languages, or who are conversant in their cultures.

Since language is often the key factor in locating overseas, ask the following questions: What is your existing and future customer base, in what languages do they prefer to communicate and where can you find enough reps to speak them?

Some reps are native speakers of certain languages; others have learned them. A native speaker uses a language in daily conversation, and understands the dialects and nuances.

Customers who speak certain languages, like French, French Canadian, German, Japanese, Korean and Thai, are particularly sensitive to the lack of nuances from non-native speakers.

"If the technical support is only business-to-business, then most likely the IT user will be able to handle English, which seems to be the lingua franca of the IT world," explains teleservices consultant Philip Cohen (Skelleftea, Sweden). "But if the tech support is more for the private consumer, I do not think English would suffice."

Adds Cohen: "A lot of people have difficulty with the language even if they can get by on vacation: by definition anyone calling a tech help desk has some sort of technical problem, and it may require a technical answer to solve it — hard enough to understand in one's own language, let alone in another lingo."

Dennis Smith, president of PacTac Advisors (New York, NY), which works with companies in Asia Pacific, finds that many software companies are comfortable providing only English language support. But as more and more have their software localized into Asian languages, or if they sell to consumers, they need language skills.

A tactic in some support centers is to enable customers to reach native speakers only when they have challenging questions and the first people they reach cannot help them.

"There are call centers that cover multiple languages in other countries, just as you have teleservices companies in the US which say they cover a dozen languages from their US centers," says Smith. "But there's a difference between accidentally having a couple of Korean speakers in Maryland, for example, and needing 30 of them to provide three-shift support at a single location in the Asia Pacific region."

According to Smith, Australia and Singapore attract nearly all the multilingual call centers for the region. Why? The lifestyle in these two countries continues to attract skilled reps who communicate in all the principal languages of the Asia-Pacific region.

"Most companies want those to be native language skills," says Smith. "In my experience, the only companies which will accept second language speakers are those running large centers which are not support desk operations."

It may be easy for reps who speak multiple languages to carry on conversations in three languages; but, when they have to handle technical conversations, it's more difficult. Yet, as more support occurs on-line, and the rep must both speak *and write* the language in technical terms, being able to speak a language isn't enough. That's why Smith predicts that on-line support will increase the demand for native-language speakers.

In assessing a location, Smith advises finding out what experience local recruiting firms have with recruiting staff, with identifying language skills, and with measuring support skills. It's also worthwhile to find out how flexible local educational institutions are in training people for support: Are they willing to assist, and have they assisted anyone in the past?

If you plan to outsource support overseas, be selective when choosing a service bureau. Support calls are less scripted than many other types of calls to call centers, so customer service skills become more important. The frustrated callers who can't get their software to work require more conversational skills and understanding of Americanized English, in addition to knowledge of how to solve problems. These are often high-value customers, for whom a company is willing to tolerate the cost of longer calls and more handholding.

"All of this makes outsourcing of multilingual support more sensitive, and therefore more costly," explains Smith. "It also increases the reliance on the teleservices company delivering what it promises. The experienced companies can handle this."

Seven

CHAPTER SEVEN

SUPPORT CENTER

TECHNOLOGY

Support Center Technology

CHAPTER Seven

Customer support centers are, on the surface, no different than many other types of call centers in their application of call and contact handling technology. They must take or make the calls, receive and transmit e-mails, route inbound contacts to available reps and to special groups of reps trained and qualified to support certain products, and maybe dial-up outbound calls.

Because customers requiring support use whatever channel is convenient for them at the time, support centers must be ready to handle calls, e-mail, chat and faxes. That means you will need tools such as automatic call distributors (ACDs), call and contact routing software, interactive voice response (IVR) units, middleware and, for sophisticated support, remote diagnostic and healing software. You also need tools to track where customers have been on your IVR and Web self-service sites so that your reps won't waste their time and yours asking needless questions like 'did you do X' if the customer already answered that question via the IVR interaction.

We strongly recommend that you read The Call Center Handbook, by Keith Dawson, A Practical Guide to Call Center Technology by Andrew Waite and The Telephony Book by Jane Laino, to give you a firm background in call center technologies. Co-author Brendan Read's book, Designing the Best Call Center for Your Business gives some very savvy and unique tips when deciding on and sourcing technology. If you're a true technoid at heart delve into other CMP books such as The Computer Telephony Encyclopedia by Richard 'Zippy' Grigonis that thoroughly explains the technological wizardry of telecom, data and networking hardware and software behind the beige metal curtain.

Where customer support centers differ from other call centers is in the volume, length, complexity and urgency of the calls and contacts. If your company does not make or market high-tech goods and services like software and Internet services, and instead, sells mainly to consumers and necessary repairs are done at depots and by field service, support will typically make up only a small percentage of contacts. But if your company offers high-tech products or communications and sells to businesses, support usually *is* service. The call and contact lengths tend to be longer, and the urgency is greater.

Remember, the customer has a problem with your product and because it isn't working it is causing problems *for them*. It isn't a call about how late you're open or if you have wax sunshades in stock.

To solve problems, no matter how much of your customer service is devoted to support, usually requires tools to track and manage their resolution, and to enable it. That also requires knowledge delivery software and systems, the principles of which are discussed in chapter 10.

Another area where they differ is that support centers make a much more limited use of outbound calling than other call center types. The only times outbound is generally used is for a callback, either from a voicemail or e-mail message left by the customer or from them clicking the 'call me' button on a Web site or IVR touchpad number or in extreme cases, a serious product flaw or recall.

That means you won't need sophisticated tools like predictive dialers for support centers but you may need preview dialing, where the customer's contact information and file appear on screen, to save reps time in making the calls. If you have a mass problem with simple instructions that requires a lot of outbound calling, like a product recall, you are probably better off outsourcing this; most outsourcers have predictive dialers. No sense wasting the time and attention spans of a $15/hour level-one rep doing the job of a $9/hour outsourcer's agent, or throwing away money on equipping a support center with predictive dialing tools that will be used once in a blue moon.

What customer support often makes a big use of is outbound e-mail, such as for fixes and product upgrade announcements. Notifying your customer base that there are fixes and new versions available saves them and you grief, and avoids costly and time-consuming calls.

Another technology you should avoid for large-scale support is voice mail.

While it is fine for reps in small firms or at upper levels who don't handle large numbers of calls throughout the day, it does not make sense for support reps because they devote nearly all of their time to communicating with customers who need their help right away. It also creates backlogs of messages that reps have to respond to in addition to doing their regular jobs.

Voice mail is also more annoying to a customer than e-mail. A rep can see and respond to an e-mail while on a call but if that call goes into 'voice mail jail' it lies imprisoned until the 'telewarden' keys in the command. Customers know this and often resent it because they know for certain that their problems are not being solved. They risk having to leave multiple messages over periods of hours or days.

Lastly, and seemingly by definition, customers of products and services requiring support, such as computer hardware and software and communications and their customer support centers are more likely to use e-mail and the Web than customers of products and services that do not, like cereal and toilet paper. High-tech customers live and breathe this stuff, and expect your support center and its staff to do likewise. They won't tolerate slow, ancient systems and technology. They will prefer e-mail to outbound calls. They will demand the ability to fix most of their problems on-line or download bug fixes. And, if and when they need a rep, that person had better know what they're talking about and can fix their problem with something more than a manual that the customers can download themselves.

In many cases you're better off having serious techie customers talking or chatting to machines — they speak the same language. Many would have no more idea how to talk to a human than if they had to communicate to the week-old mold at the bottom of their 'java delivery device' a.k.a. coffee mug. (The less charitable amongst us also would claim that there is little difference between the two life forms.) But because your customers are human you need to have humans on the phone. And you can train the less morphed technoids to communicate like a human — at least while supporting your customers (see chapter 8).

Key Support Tools

There are several important tools and technologies that you must consider deploying in your support center to carry out these three principles. Some of them are for internal support as well as customer support while others are used in call centers, in general.

Trouble Tickets/Problem Management

The most important tool in the support center, arguably more vital than the phone, is the trouble ticket. The trouble ticket is the record of the customer's problem: who is the customer, their contact information, the product they have the problem with, a description of the problem, and action on it. Support problems are often known as 'cases' or 'issues.' All trouble tickets have tracking or case numbers.

In the old days a trouble ticket was just that, a paper ticket attached to a piece of machinery. Now trouble tickets are e-mails and more recently on-line forms, filled out by customers or by reps when they take the call.

On-line forms are superior to e-mail forms because e-mail often gets lost in the barrages reps receive. Sometimes they are not stored. With on-line forms

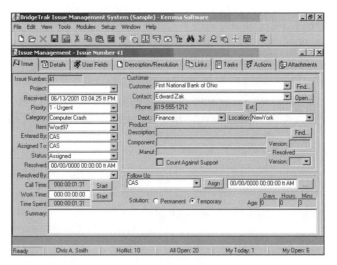

Kemma Software's BridgeTrak Issue Management screen is an excellent illustration of a software-based trouble ticket. It shows the customer's name and contact person when the trouble was reported, the name of the rep handling it and actions taken, in easy to access tabs.

connected to your Web site there is no excuse for reps not to see them. Customer and internal support centers alike use these tools.

With on-line trouble tickets customer can also update them. That's handy for the reps so that when the customers call or e-mail back they know the latest developments and can help them more effectively.

Problem management software creates, assigns, updates and closes issues and cases. They track tickets to their resolution, including escalation to upper level reps or to field service. Another key function of problem management software is to ensure reps are assisting customers within the timeframes defined in their service level agreement (SLAs). SLAs are written contracts that stipulate that reps must respond within a certain number of minutes, hours or days.

It's a good idea to set internal timeframes for escalations that are sooner than customers' SLAs. We cover SLAs and other topics related to charging for support in chapter 3.

There are several problem management products and vendors. For example, SupportWizard, a division of Integral Solutions Corporation (Redwood City, CA) automates the processing of new tickets with fully customizable workflows, automatic e-mail response and notification, user defined escalation rules, customizable java chart reporting, and sophisticated control over access.

SupportWizard has a knowledge base with dynamic FAQs to help your customers find immediate answers. It comes with a fully licensed MySQL database and permits complete customization of fields, choices, look and feel (including graphics, templates, fonts, and footnotes), and user groups and permissions.

You can also use SupportWizard to send bulk e-mails to selected people in the user table or to authors of selected tickets (based on a search or on manual selection). The e-mail may be sent from any 'From' address and cc'd to an unlimited number of recipients. You can use this capability to send surveys to customers, to notify them about new products/upgrades, or for any other purpose.

Customer Tracking

Customers seek support through several channels, including e-mail, phone and through IVR and Web self-service. Just as trouble tickets track support issues so must you capture and track where the customers have been on your IVR and Web self-service platforms so that if and when they call or e-mail, the reps can better solve the problem. If they tried A and B, that saves the rep from suggesting those options, and from hearing "look, moron, I've done that and it's still not working!! Instead your auditory-fireproofed expert will suggest C.

Rick Kilton, RWK Enterprises so aptly points out in other chapters, knowing what customers have tried and the errors they made in doing so enables you to let them know how they could have navigated the site easier. Knowing how they got around the site also enables you to find ways of improving it.

More software is becoming available to allow support centers to gather customers' on-line actions and present the information in a screen pop to the reps' desktops such as Cosmocom's (Melville, NY) Universe platform, Siebel's (San Mateo, CA) Call Center 7 and Nortel Network's Symposium Web Center Portal (Research Triangle Park, NC). These products and others like them let you establish rules that define how long a customer is on your Web site before it starts capturing screens.

In this same vein, your reps can also help customers complete electronic forms on your Web site through the use of software such as Hipbone Inc.'s (San Carlos, CA) Synetry suite, Com-Network's (Paris, France) Call Operator suite, and NetDIVE Inc.'s (San Francisco, CA) Enterprise suite.

Problem Resolution

In the beginning problems were resolved by reading manuals, scratching calculations and drawings, and one's head to stimulate the brain cells and poking testing equipment into the box (making sure it was unplugged, of course). But many times these exercises were the support center equivalent of reinventing the wheel; imagine a whole room of people scratching their heads. Sweet dreams for the Rogaine makers.

Studies show that most of the problems that support centers deal with have the same answers. Problem resolution technology, also known as "canned brains" provides pre-packaged and updatable answers to problems contained or accessed in a knowledge base, which in most case is a specialized database (more about this in Chapter 10 on knowledge management) or in databases, files and Web sites in your business's or in partners' or suppliers' databases, files and sites.

The software saves the company and its customers time and money; customers can access these answers through low-cost Web self-service that leaves the reps, usually level-one, to hand hold customers and to resolve the issue if the answers are not found in the knowledge base.

There are a host of vendors who provide problem resolution software, with a variety of features. For example, eGain's (Sunnyvale, CA) Knowledge Gateway enables reps and customers to find consistent, accurate answers through a wide variety of sources, including Lotus Notes and Microsoft Site Server indexes of ODBC databases, Microsoft Office documents, HTML documents, and e-mail files.

The Knowledge Gateway lets them tap into existing, less-structured corporate and external information that can answer an inquiry. It extends the intelligent searching capabilities and intuitive, conversation-driven access of the Knowledge Gateway to your unstructured information. This intelligent gateway uses groupings of related ideas, known as "taxonomies," that search concepts and not just key words — helping users refine their search to quickly find the exact information they need.

This type of software also helps control and limit escalations to $25/hour level-two and higher reps. The FrontRange Solutions (Colorado Springs, CO) Heat software's Business Processes Automation Module (BPAM) enables you to create real-language escalations and automation rules using best practice guidelines, then check calls against these rules and receive an alert when rules are being met.

To enable a knowledge base to work you must construct it. Problem resolution vendors such as ServiceWare (Oakmont, PA) have such toolkits. The eService Architect provides your knowledge engineers with browser-based tools to build, create and maintain a robust knowledge base. The application enables your engineers to review new knowledge submissions, provide style and technical reviews, and determine which knowledge self-help users can view and which should remain internal for analyst use.

The software has remote browser-based authoring, a WYSIWYG editor so that knowledge authors can easily create and maintain knowledge in HTML format and a knowledge quality scheme — allows for 'promoting' knowledge

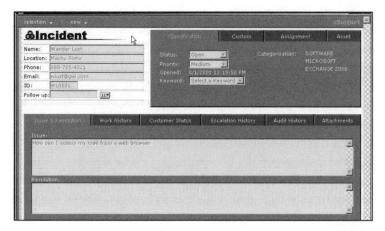

GWI's c.Suport software lets reps assign a numerical value to customers' complaints to indicate severity. After each case, reps can save a history of their correspondence with the customer, including notes on phone calls and e-mail mes-

from pending to active after proceeding through QA steps that you determine.

Call Center magazine Associate Editor Lee Hollman reports that some vendors offer customer support and internal support versions of the same software. GWI's (Vancouver, WA) c.Support Version 8.0 was designed for reps who handle general customer support calls and c.Support IT for help desk technicians. You install c.Support on a Lotus Notes or Domino server; and c.Support IT runs with Microsoft Exchange 2000.

c.Support IT enables customers to request assistance by filling out on-line forms, participating in live chat with reps and searching an on-line database from your company's Web site. Reps can also create and save trouble tickets for customers. Both versions of the software also let you generate reports that contain information including the number of open calls and the priority level that reps assign to each call.

"The biggest difference between our IT version and our customer support version is how we name things within the [software]," says Daren Nelson, ceo of GWI. "The business processes are unbelievably similar in 95% of the cases." For example, customer support reps might refer to customers as customers, while help desk technicians refer to them as end users. While c.Support and c.Support IT cater to differences in user jargon, they share basic features.

Applied Innovation Management (Las Vegas, NV) has two products: HelpDesk Expert for Customer Support, which provides agents with multimedia communication; and HelpDesk Expert for IT Support, which offers tech support to call center agents and to other employees.

The HelpDesk product line helps on-line customers through a Web browser. The software scans customers' e-mails for keywords to route them to qualified reps. Reps can also attach files, such as Adobe Acrobat documents, to e-mails they send customers. For example, if asked about a specific product, a rep can send a brochure as an Adobe Acrobat file. The software also lets customers search an on-line knowledge base from your Web site and download software or multimedia presentations to resolve issues.

Additional modules for the help desk software let reps use a whiteboard to draw diagrams for customers; track the status of customers' service agreements and contracts; and provide on-line surveys to measure customer satisfaction.

Applied Innovation Management also introduced Liz, software which lets reps listen to customers' recorded phone messages as MP3 files using a PC. The app automatically assigns customers trouble tickets that they can view from your Web site. Reps can use Liz with HelpDesk Expert for Customer Support to play back MP3 files and to view customers' histories through a Web browser.

Using HelpDesk Expert for IT Support, reps can submit trouble tickets to your IT staff by e-mail, check the status of trouble tickets and search a knowledge base to solve problems.

Andy Bestwick, sales manager for Applied Innovation Management, says that differences between the two versions of HelpDesk Expert mostly pertain to jargon. But he adds that HelpDesk Expert for Customer Support necessarily includes more features than the IT version of the software, since customer support reps have to help customers with a broader range of requests and issues.

Call/Contact Distribution

The easiest way to set up support is informally. A call comes into your company through the PBX, the caller claims the recycled player piano scroll masquerading as computer software has gummed up their hard drive, the receptionists roll their eyes for the hundredth time that morning, and transfer the call to the unlucky engineer who missed the 10 am espresso bar rush. Or if you use a key system the reception puts the call on hold and yells "Alexander, Line 666!" If you employ a direct inward dial (DID) system Alexander's phone rings and he can then roll his eyes for the hundredth time.

But as the business grows, support volume increases to the point where you need tens of people to handle support calls and contacts. That's when you need to route them efficiently, with automatic call distribution to the first available rep.

The automatic call distribution device is the nerve center of any call center; it routes all calls that come into the center and manages information associated with those calls.

The device you opt to use for automatic call distribution is typically an automatic call distributor or ACD. But automatic call distribution can be carried out by a wide variety of hardware and software. For example, you could use a type of telephone switch with highly specialized features and robust call processing capabilities such as those found in large call centers.

The ACD, which can be hardware or software or both, connects callers with the reps usually based on the first available rep. Because the ACD is the channel that your customers' calls flow through, there is a flood of consistent, accurate and real-time data such as call volume, queue size and speed of answer it can provide that enables a support center management team to efficiently and effectively act.

Vendors also make monitoring and recording devices and workforce management software that connect to ACDs to gather information from its call and contact stream. Then to help regulate call flow, IVR systems, hardware or software are attached to the ACD. IVRs can answer some of the calls directly. There's more about IVRs later.

Typically support centers that have 30 or more seats benefit from ACDs. There are many companies, i.e. Alcatel, Aspect, Avaya, Cisco, Intecom, Nortel and Rockwell to name a few, that sell large, expensive ACD systems. If you have a popular product, like consumer software, or Internet services, and get many calls that require big pools of reps, or you are an outsourcer with such clients then your support center cannot function without one of these monsters.

But if you have less than 30 seats there are PC-based ACDs, PBXs and key systems with ACD functions, and even standalone ACDs that you may find efficient and effective. There are ACDs designed specifically for a small center such as Cintech Solutions' (Cincinnati, Ohio) Prelude and Cinphony ACDs.

Let's not forget one other automatic call distribution system, the communications server. A communications server can be cost-effective for the small support center because it can combine a number of call processing and voice processing features into a single system. A good example of a communications server is one offered by one of the best-known manufacturers of communications servers, Interactive Intelligence (Indianapolis, IN). Its Enterprise Interaction Center (EIC) includes a PC-based phone switch, an ACD, an IVR system, a call monitoring system and computer telephony software. With EIC, you can apply routing rules to phone calls and to on-line communication like e-mail.

Product Features	EIC	CIC
PBX	Yes	Yes
ACD	Yes*	Yes
IVR	No	Yes
Unified messaging	Yes	Yes
Skills-based routing	No	Yes
Screen pops	DDE only	DDE & COM api
Fax server	Yes	Yes
Auto attendant	Yes	Yes
Reporting	Optional	Yes
Presence management (real-time status information)	Yes	Yes
Interaction Designer (handlers development)	No	Yes
Interaction Administrator	Yes**	Yes
Remote agent capability	Yes	Yes
Web Features		
e-mail queuing	No	Yes
Web callback	Yes	Yes
Web chat	Yes***	Yes
All-in-One PBX/ACD/ unified messaging/ fax server, automated attendant (no custom handlers for EIC)	Yes	Yes
Customizations to any feature above	No	YES!
Add-on Products	**EIC**	**CIC**
e-FAQ™	Yes	Yes
Interaction Dialer™	No	Yes
Interaction Director™	No	Yes
Interaction Recorder™	Yes****	Yes
Wireless Interaction Client™	Yes	Yes

When shopping for products like ACDs and routers vendors often offer equally robust 'lite' and full-featured products. That way you buy only what you need.

Routing Software

Once you have your automatic call distribution system in place, you must determine what types of calls your support center receives most often, and which reps are best equipped to handle them (see chapter 2). With these parameters in hand, you can set up your routing rules.

The basis of all routing is a queue. A queue is a line. As at a movie theater, where the first person on the line is the first person in the theater, call centers often handle calls in the order in which they receive them.

To vary from this simple queue set-up requires what's referred to as queue management. And, for a support center to operate efficiently and effectively there must be good queue management. If not, customers are left holding for long periods of time, and when they finally reach a rep they spend valuable time venting their irritation and frustration. This not only can cause customer attrition, irate customers also result in a group of stressed out reps, which, in turn, affect the center's attrition rates. A poorly managed queue also results in varying degrees of overstaffing. All of which can add substantially to the center's operating costs.

Routing software is a good queue management tool and with the use of specific routing rules, a support center manager can establish exceptions to help bring order to a queue. For example, Jim speaks English and Japanese fluently. Cliff and Rich are experts on 'Cosmo2002,' the company's newest product, which comes with a bevy of new bells and whistles, while Madeline is a whiz on its older sibling, 'Cosmo2000'. There are also experienced reps that know all there is to know about the company's legacy products, which still have a sizable install-base.

But, how do you account for all of these variables? How do you get your customers in touch with the reps who can best serve them and still meet your service goals? And how do you factor all this in without manually switching reps from group to group and also avoid over burdening agents who have multiple skills? Situations like these clearly demonstrate why call routing software is one of the most important components of your support center.

A routing system, whether it resides within your phone switch or runs as a distinct ACD, enables you to establish exceptions to the queue. A phone switch often comes with proprietary routing software that runs on a server connected to the switch or ACD. With this software, you can indicate which support reps are at which phone numbers or extensions, and which groups reps belong to.

Call routing software lets you expand on your switch's or ACD's capabilities by directing calls to reps based on each rep's skills *and* the customers' preferences, among other criteria. Most routing systems allow you to qualify each phone call before sending it to a rep. Instead of assigning priorities to callers or to reps' skills, you can program the system to take into consideration several factors before the call is routed to reps.

For instance, the system can route phone calls based on your customers' profiles. By integrating with computer telephony software and/or your IVR system, the routing software can identify customers and send their calls to appropriate reps.

You can set up different routing priorities and objectives for different types of callers. For example, high-priority customers like those who purchased your Plutonium service plan that promises to deliver a field service technician to their doorstep by Harrier jump-jet if the support rep can't debug your BugMaster uninstaller in 20 minutes or less. Or when you set up a rep's profile, you can factor in their preferences as well as one or more skills they might possess. For example, if a rep speaks American, Australian, Anglo-Canadian (eh?), Irish and NooYawkese as well as English but would rather take calls from dose that know where toidy-toid and toid is and don't have a problem with that, then you can indicate that the rep's main preference is to answer calls in NooYawkese.

When a call comes into the center, the routing system should consider the customer, the number of available reps, their skills and your service level requirements. For example, let's say your center has three callers waiting on hold and the customers who is third in queue is the only one among the

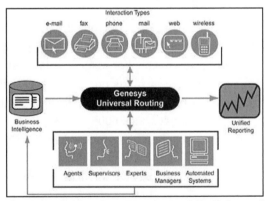

Genesys Lab's (San Francisco, CA) Universal Queue2 software routes interactions of all media types through a single, integrated media queue. Interactions are distributed according to company-defined routing strategies to the most appropriate call center resource. That lets you leverage all customer knowledge-enterprise-wide-to deliver a more consistent, more personal interaction and apply that customer knowledge to make each interaction more effective.

three who is a high-priority 'Plutonium' customer. If the routing system estimates that you will not meet your service level for that customer unless it routes that customer to a rep ahead of the other two callers, it will do so.

With most routing systems, you can also assign 'reserve' reps, i.e. reps who are available to answer calls if reps in other groups experience unusually high call volumes. Many of these systems can also monitor the amount of time that reps spend on the phone and evenly distribute calls throughout your center, routing calls to reps that have spent the least amount of time on the phone instead of directing callers to the first available rep. For example, a variety of routing software can recognize the difference between an agent who has remained on a single call for 30 minutes and an agent who, during the same time, was on a break for 20 minutes and on a call for only ten minutes.

Increasingly, routing software must handle e-mail and chat as well as voice. As with call routing software, multimedia routing software lets you apply what you know about customers to build rules for directing calls and on-line messages to reps.

And just as call routing software gives you a choice of switches through which you direct calls, the multimedia counterparts of these products allow you to establish routing rules regardless of the software you use to keep track of information about customers.

Even with the aid of reporting and workforce management tools, it's difficult to predict how many customers will call and what they will need. It doesn't get easier with the emergence of on-line customers. Although companies no longer make millions by promoting themselves as dot-coms, the legacy of the Internet boom is that customers believe they ought to find answers to questions on Web sites, whether they look them up themselves or correspond electronically with reps.

One way to maintain control of expectations, at least within your support center, is to use a single routing engine to direct all communication from customers. Multimedia routing software is often available as a set of modules that correspond to different types of communication. The logic behind the modules is that it's unlikely that you or your colleagues in other centers will use all of them at once.

Beyond sending e-mail messages and calls to reps, multimedia routing software enables reps to receive live text messages from customers, as well as on-line forms from visitors to your Web site who request calls at a later date or time. Multimedia routing software also let reps and customers view the same Web pages together.

Whether they send e-mail or live text messages to reps, most multimedia routing software products include tools that let you create libraries of responses for live responses or automated replies. When you build routing rules, you determine which types of messages receive automated replies and which go to reps with suggestions for replies.

These libraries need not be as extensive as the knowledge bases we discuss in chapter 10, although in an ideal world, your response library should be able to serve as your knowledge base if you don't already have one. And if you do have a knowledge base, your routing software should be able to gather items from it.

Unfortunately, this is not usually the case with multimedia routing software products because they do not automatically integrate with knowledge management tools. For now, one way to save your company the time of building the same knowledge bases over again is to consider software that does not direct calls to reps, but does accommodate on-line messages.

Routing software vendors offer products that use many different approaches in their effort to provide support centers (and call centers in general) a routing system that will fit their specific needs. Let's take a look at a few of these products.

Nortel Networks offers its Symposium family of products, which takes the approach of associating skills with agents instead of associating agents with skills. Symposium's skill set assignment matrix gives supervisors more control over call routing and allows them to make changes to routing rules on the fly. Then through real-time displays and historical reports, you can keep track of all of your customers' communication with reps.

Siemens (Boca Raton, FL), a switch vendor, has an interesting approach to call routing. Its ResumeRouting family of software can route, in addition to phone calls, IP telephony calls, e-mail messages and requests for on-line video sessions using software such as Microsoft's NetMeeting. ResumeRouting products allow you to build profiles for each rep in your center. In these profiles, you include reps' skills, their skill levels and preferences. These rules enable your center to route calls to reps whose skill levels and preferences match up with the type of rep the customer wishes to reach, based on the caller's selection from an IVR menu or based on the specific toll-free number the caller dials.

Rockwell Electronic Commerce's (Wood Dale, IL) Contact Integration Management (CIM) software let's support centers give certain customers higher priorities than others. Aside from routing that reflects reps' skills, this software lets you route calls based on customers' profiles. CIM also provides supervisors with real-time and historical information about reps' productivity.

If you are looking for software that is switch independent, you might want to consider Genesys' Enterprise Routing software, which works with most major switches and routes phone calls, e-mail messages, voice-over-IP calls and Web callback requests. The software lets you create rules for routing based on customers' profiles, the numbers customers dial to reach your center, the numbers customers call from or information customers enter through your IVR system. The software factors in call volumes and reps' availability. You can assign reps to different groups and establish proficiency levels for each rep with regard to one or more skills.

But, what if the majority of calls to your center are from customers with varying levels of service level agreements? In such a case you might want routing software that lets you set up criteria for routing calls from particular groups of callers. That's exactly what Xantel's (Phoenix, AZ) Connex 80/20 PCD products do. Connex 80/20 PCD identifies callers using ANI, DNIS or account numbers customers enter from your IVR system. The software employs what

Monitoring

Quality customer support is vital to retaining and attracting customers, and if you charge for support, in ensuring that your revenue stream continues and grows.

One way to ensure that is through monitoring reps, on the phone and online, through screen captures. There are many tools to accomplish this, developed by vendors such as Envision Telephony, Funk Software and Witness Systems.

Recordings of calls and screen captures, as well as feedback from customers, are essential to your quality assurance efforts. Use them to evaluate and train reps.

But before you make any plans to monitor calls, check on current and pending legislation. Laws vary from state to state as to whether reps and customers have to give prior consent to recordings of their conversations. You can usually find this information by viewing the Web sites of the states where your company does business.

Xantel refers to as ServiceTiers, which let you define rules for certain customers. Consequently, you can set up criteria for directing customers to the same agents or the same groups of agents every time they contact your center.

Interactive Intelligence's EIC can route phone calls, direct on-line communication to reps and enable reps to push Web pages to customers because with each incoming interaction EIC performs a series of calculations to determine the best match for the customer, the rep and the interaction. For instance, EIC considers reps' skills, the length of idle time per rep, the skills reps need to handle an interaction, the priority level of the call or on-line request, and the length of time the request is in queue. You assign weights to determine the importance of each routing criterion.

By factoring in information gathered from your IVR system or a Web page your customers visit, you can assign some customers a higher priority than others. You can also restrict reps to certain types of interactions, such as phone calls or on-line requests since EIC lets you assign reps to multiple groups based on skill sets, set levels of proficiency and preferences.

Here's how it would work. Let's say that Jim, the aforementioned rep who demonstrated high proficiency in speaking French, doesn't have much of a preference for using this skill in dealing with callers, and Deborah, another agent of equal skill, has both high proficiency and interest in communicating with French-speaking callers, given the choice of where to direct a French-speaking caller, EIC would opt for Deborah over Jim.

IVR

IVR is one of the most versatile technologies call centers and support centers can deploy. As the above suggests, IVR is a vital routing tool. Customers key in, or with speech recognition software say their account number, password or platform type and they can get connected to the right rep.

IVR also saves money. As support volume grows it becomes more costly to handle it. Having problems solved without having reps solving them saves money. Automated support: e-mail, IVR and Web costs pennies per transaction compared with dollars for live reps (see chapter XX).

Also, unlike live reps it doesn't matter where your automated support is located, as long as the server box has fat pipes and redundancy, and you have redundant servers in at least one other location either owned or outsourced, in case one goes down. Unlike your 'loyal' employees you don't have to worry about an IVR port walking out of the computer room to the one next door for eight bits more per hour. As long as you keep them cool, comfy and juiced, and you don't let unauthorized personnel play with them or 'borrow' boards for their Frankenstein experiments they should be fine.

You can offer two ways for customers to get help without communicating with reps. One, which we focus on in this section, is by using an IVR system. The other way, as we explain in chapter 10, is to allow customers to refer to on-line knowledge bases.

> ❖ **NOTE:** We have deliberately stayed away from Web self-service technology, Web site hosting and carrier selection and call accounting in this book. Support represents only a portion of Web site functionality and phone demand; the Web teams and telecom managers have other demands that they must meet, like marketing and sales, which are big users of these media. Companies therefore choose Web software and carriers and carrier-related products based on the needs of all internal users, not just the support center and team.

IVRs enable callers to look up information about a company and its products and services by selecting choices from a menu of options from their touchtone phones. They can perform simple tasks like password reset. Support centers use IVR systems because they can provide their customers with information quickly and economically. Besides, customers, in general, have easier access to a phone than to a computer.

On the phone, customers can often choose whether they want to look up information by themselves from an IVR system or transfer out of an IVR system to reach live agents. Analogous options are not always available from Web sites.

Keep in mind that support centers must adopt the customers' perspective to find the right system to use, i.e. you need to understand what basic problems customers have and what they are trying to solve. Many managers find that if not managed carefully, IVR can be a source of frustration for their customers, especially when IVR systems do not respond as callers expect.

Also, IVR systems aren't always necessary for small support centers. If your center has less than 20 reps, do a business study prior to embarking on your journey into voice response and processing systems. Find out what portion of calls can be farmed out to an automated facility. If the results indicate some type of automated assistance by phone is advisable, you will find that there are a wide variety of products to choose from.

And when purchasing an IVR system, it's best to keep in mind that you may eventually have to upgrade not one but multiple components on the same system. Therefore, you need to be aware of what you'll need tomorrow as well as today.

The more sophisticated your support center, the more robust features needed. For example, you might want an IVR system that has speech recognition capability and that can work in conjunction with call recording and monitoring systems.

Speech recognition has some value in support centers. You can provide speech rec for certain tasks, like stating issues with specific components of stereo or computer systems, where callers tend to prefer describing problems in words to figuring out what digits to press on their phones.

Letting customers talk to a machine is more natural than asking them to press buttons. But, if customers are excited, it may not pick up the dialogue, which might not be a bad thing!

IVR vendors offer many advanced options, such as databases that collect information about system use for service analysis or for input into other company systems. Such as voice forms that provide a structured mechanism for collecting data, e.g. information for warranty verification and credit card authorization services that allow customers to input card numbers and payment amount and validate the entire transaction.

Few support centers make the most of their IVR systems. Many centers use these systems to allow customers the option of selecting digits on touchtone phones to reach a particular rep or group of reps. That's fine for routing, especially if you enable customers to enter their account numbers or numbers associated with their trouble tickets, and if you display these numbers on the reps' computer screens. In real life, however, customers frequently find a way to transfer from the IVR system to reach a live rep, if such a choice is available.

The IVR vendor community is both very specialized and highly fragmented. For example, there are software tool kit vendors. Within this group are vendors such as Artisoft (Tucson, AZ); Dialogic (Parsippany, NJ), a division of Intel; and Pronexus (Kanata, Ontario,Canada). The software tool kit approach requires the buyer to acquire and integrate software and hardware, develop the application, install, test, train users and maintain the system. The payback is a system with low out-of-pocket costs and a low cost of replication across multiple sites.

Then there are generalist IVR vendors, which include InterVoice-Brite (Heathrow, FL); Edify (Santa Clara, CA); and Lucent (Basking Ridge, NJ).

Let's look at a few of the vendors who offer IVR products that work well in small to mid-sized support centers.

Some IVR systems, like the IntelliSystem (Reno, NV), offer more than simple automated responses. The IntelliSystem is an IVR system that callers can use to identify possible solutions to technical problems by asking callers what products or problems they are calling about. Once the customer provides the information, the system refers to a knowledge base, which enables it to offer the customer a specific series of instructions about how to handle the situation.

But at all times, the IntelliSystem allows the caller to transfer out of the IVR system to speak to live reps. It also lets callers save an IVR session so that when they call back, they can restart an IVR session at the point where they left off. The IntelliSystem then generates reports that tell you how many people received help and how many customers abandoned.

When you offer IVR service you must provide fast, competent automated service. Your center's success depends not only on being able to cost effectively handle a high volume of incoming calls, but also providing consistent, excellent service.

Enter Edify Electronic Workforce (Santa Clara, CA), which has the capability to act as an IVR unit. Not limited to the simple inquires handled by traditional IVR systems, Edify's product can automate complex, detailed, and varying interactions. Integrated technologies like natural language speech recognition, voice authentication, and advanced text-to-speech allow customers to automatically interact with your company in ways that were not previously possible.

The Electronic Workforce is an ideal way to automate interactive transactions and processes. Offering the full feature set of traditional IVR products, Workforce Objects are preprogrammed to receive and interpret standard touchtone or speech commands, dynamically manage menu choices and options, assemble and communicate responses in real-time and transfer calls.

!hey inc.'s (North Andover, MA) IVR system is an on-site voice processing system that manages multiple self-service applications and can give your support center complete control over all its incoming calls even in a blizzard. The system manages multiple self-service applications on a 7x24 schedule. The system integrates business rules and IVR to efficiently route inbound calls through self-service menus or to designated live customer support personnel.

The product's monitoring and reporting functions continuously track customer information and measure the cost and service effectiveness of IVR applications. The IVR then retrieves information about the caller and attaches it to the call; the call-data package is routed to a support rep and provides the basis for an informed conversation about the customer's specific product configuration.

Captaris (Kirkland WA) offers innovative call management and interactive voice response that work well in the support center environment. This company's products can assist you in improving customer satisfaction and increasing customer retention through self-service options, fax response, and call qualification.

The product's Desktop Call Manager and Automated Agent features can help manage overwhelming call volumes, regulate call distribution to reps in specific groups, increase your reps' productivity, and automate customer interactions without the necessity of live intervention. For example, Desktop Call Manager can deliver skills-based call routing combined with screen pop functionality, which provides the most efficient call management solution for today's small support centers.

CTI Middleware: Connecting IVR with Routing

IVRs and ACDs are great tools. But to deliver great support you must make the sum greater than the parts by linking these tools with the customers and the customers' data, i.e. computer-telephone integration (CTI). That way when the call

IVR and FOD (Fax On Demand)

IVR and FOD provide information to a customer by turning the touchtone pad of a telephone into a keyboard. As a result, many standard queries formerly answered by reps can be handled by an IVR/FOD combo, freeing reps for more complex, value add work. FOD is a lifeline for ISPs when the Internet connections crash, which inevitably jams the phone lines.

Here's how it works. If customers have a question about a product or service, they call a number and either enter a code or choose from a list of documents to be faxed back to them. This works very well if they already know what you are looking for and just want details. Let's say that they want information about a color printer and have the model number, they could call the support center enter the printer's model number and get detailed information faxed to them.

reaches the rep all the information about that customer's product and experience. For example, they originally bought Version 0609, upgraded to 0906, but when they installed 0909 everything crapped out. The reps have pasted on a stiff upper lip, the adrenalin is pumping because it is 0900 and their head is about to be figuratively ripped off.

Enter CTI middleware. Here's how it works. A customer initially reaches an IVR system that then prompts the customer to enter identifying information such as an account number before allowing the customer to speak with a rep. This software retrieves this data, and uses it to locate a record of a customer with the same account number.

Before the phone switch routes the customer's call to a particular rep, the software verifies it can find a record for the customer. If so, the middleware triggers whatever software the rep uses to view information about customers while the call routing software directs the call to the rep. The caller's record then appears on the rep's computer.

Many vendors use the term "screen pop" to summarize the scenario we've just described. Examples of middleware include CallPath from Genesys and modules that are included with Network Associates" (Santa Clara, CA) Magic Total Service Desk Suite and U.S. Infotel Corporation's (Oklahoma City, OK) Odyssey Call Center System.

To illustrate how screen pops enable your center to route calls based on what your support center knows about customers, let's say callers to your center identify themselves by reaching an IVR system; they then key in account numbers from their touchtone phones and select the groups of reps they want to reach.

With this information, you can set up rules that move certain customers ahead of the queue, either because customers' reasons for requesting support are more urgent or because your company's records show that certain customers, by paying extra for support, should receive higher priority than others. And they won't nearly be as furious as they would be if they have repeat and re-enter the same !#$%&*() information from when they called and contacted your firm before.

Remote Servicing/Diagnostics/Self-Healing

The best rep, live or automated, may be the little person who lives in your customer's machine, program or network, or who can be sent on their way with a microvolt jolt to their virtual overalls down the copper or fiber optic pipe or through the ether to find out what the problem is and fix it (see chapter 10).

This is the essence of remote diagnostics, servicing and healing software, either hosted and managed by the support center or pre-loaded or post-loaded into the equipment.

These tools can upgrade and evaluate many types of devices, such as computers, electronics and even household appliances. The information gleaned from a quick peek inside a customer's seemingly dead box allows the support rep to provide a quicker and more cost-effective solution to the problem and/or alter the device's parameters to re-optimize its performance without the need of a costly 'truck roll.'

From a software standpoint the tools must be installed on the customers' devices, then from a hardware standpoint the remote diagnostics tools do their 'magic' by accessing the customers' devices through a standard phone line or high-speed connection, which connects the customers to the support center's systems through a modem. With the locations physically linked, the support center's systems can act as a terminal in accessing information located within the onboard intelligence of the customers' devices.

Together these tools can save your support center and your customers a lot of time, money and hassle. Scott Harmon, chief executive of vendor Motive Communications (Austin, TX) told *Investor's Business Daily* (October 4, 2001) that self-diagnosis and healing could cut live-rep call and contact demand by 5% to 30%. The story reported a recent Forrester Research survey that found that 62% of companies rank self-service as more important than all other customer support aspects.

But you have to be careful how you deploy these tools. You and the vendor must make sure your customers clearly know what technology they are intended to diagnose and fix.

The October 4 *Investors Business Daily* article on Motive reported that Dell had removed the Motive application because though it worked fine, users hit the Motive icon to fix software problems instead of the intended Dell hardware problems. Dell spokesperson Bryant Hilton told the newspaper that his firm has "taken the tool back to the drawing board so these tools can fix problems with configurations, driver access, corrupted files and registrations. If it can't be fixed automatically, the issue is escalated to your support desk.

Motive's tools also create profiles that prevent end customers from foisting changes that could break your software. It estimates that 13% of all problems are self-inflicted: by the end-user, your customers.

These tools are especially useful for broadband support. SupportSoft

(Redwood City, CA), a leading broadband software company, developed BroadJump, which it incorporated into its Broadband Resolution Suite to enable subscribers to quickly resolve problems by enabling automated diagnosis and resolution for both PC and network/service connectivity (including CPE modem) problems.

Hosted Systems and Software

The proliferation of the public Internet and secure private networks means that your company can use software and equipment from locations outside your support center instead of installing new products in-house.

Hosting is more of a delivery mechanism than it is a technology. Vendors provide access to their software from secure portions of their Web sites. Although multimedia routing and CRM software are the categories of products that vendors most commonly host, the jury is still out on which types of software are best for hosting.

Through the wonders of technology, vendors connect components or entire servers to your network based on, for example, how many ports of an off-site IVR system or trunks of an off-site phone switch you want to use.

Software developers typically charge monthly per-server and per-seat fees to use their products. If you're paying a vendor to host equipment or a combination of software and equipment, expect to pay fees based on the total number of calls and on-line transactions your center receives each month.

More service bureaus are hosting software, in addition to communicating with customers on their clients' behalf. The advantage for you is that if you outsource some, but not all, customer support, the reps you employ at your center can use the same software as the reps who work for the service bureau.

That's a big plus, given that service bureaus typically operate on a scale that enables them to purchase software that's more costly and powerful than any one company might be able to buy on its own. A few outsourcers also host equipment, typically IVR systems that include speech rec, besides offering your company services from live reps.

Connecting Your Support Operations

For many companies customer support can be handled simply with a customer support desk. Many of these consist of a few reps on one floor or in a corner of a floor.

Others have enough call and contact volume to merit a separate dedicated center.

But other companies' may handle customer support by outsourcers, by tele-working reps, by counter, depot and field support with some support operations spread out amongst a network of support centers worldwide.

Complicating matters is what happens when your firm acquires another and your management decides to integrate the bought-out company's support centers with yours. This is necessary because not only does your new corporate 'little brother or sister' have their own products that would require time-consuming retraining for your reps, but also introduce into the support technical ecosystem problem management/resolution platforms that don't talk to yours.

In all of these cases you need to link these support operations and personnel together. And that won't be easy. With the spiders' web of voice and data pipes that must be connected to disparate technologies that often don't talk to each other (the biggest myth in technology is 'open' standards), in various buildings and rooms, the good consultants, systems integrators and IT staff that you can find to handle this work are, if anything, underpaid for their services.

Undertaking that work is like rewiring an old office building: you really don't know what's behind the walls, you're not sure where all the wires go — many are dead ends — and something might crawl out and bite you in the sensitives. For that reason consultants recommend that as a first step you get a real good company project manager who knows your network, technology, people, and buildings, and who can work with vendors and their employees. Someone who is prepared for anything.

Network-wide Routing

To connect customer support centers there are two general routing methods — carrier-based or premises-based. Only you can determine which is best for your support operations. In carrier-based routing the carrier does the switching. In premises-based routing the owning company does it.

With carrier-based routing, the system takes a survey of the entire network at the time a call arrives at the PSTN (public switched telephone network) — before it's transferred to a rep — to determine which rep at which center is most appropriate to handle the call. In a multichannel support center, even more efficiency can be gained if the call-routing decision process takes into account the number of e-mails and chat requests that are in the queue at each center.

One of the first vendors to offer a network routing solution that could do this was Aspect Communications (San Jose, CA) through the integration of its Aspect Carrier Routing system with its Aspect Enterprise Contact Server.

It works like this: When a call arrives at the PSTN, the carrier sends a query to Aspect Carrier Routing, which in turn queries the database of the Enterprise Contact Server. Because the Enterprise Contact Server integrates with e-mail response management systems and with Aspect's Web Interaction software, the database contains real-time information on every contact, regardless of channel, at every contact center site.

The advantages of carrier-based routing are that small efficiencies gained in call routing results in overall cost savings. You also avoid incurring the capital for switching gear. The disadvantages are that the support operation is limited by the technology and the carrier's upgrade schedule. Some support center managers are also a bit wary of trusting the necessary data routing changes to the carriers.

In premises-based routing the calls and contacts come to your switch and then you set up the routing to other switches. The advantages of premises-based routing are that you have greater control over routing and the technology to do it. You, not the carrier decide when and how to upgrade it.

There is also a big cost savings in routing calls to the next available agent rather than doing a percentage allocation from the network. And if you switch carriers there is less hassle because you don't have to have the new carrier's

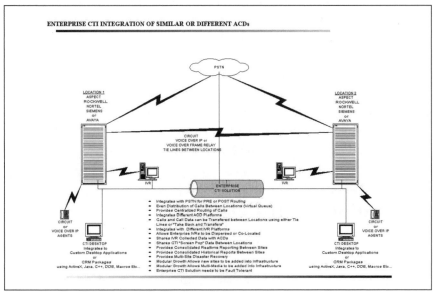

Example of call center multi-site operations.

switching and routing configured.

Premises-based routing is only cost effective if your support centers are large (200+) and in three or more locations. That's because it is more labor and capital intensive than carrier-based routing.

Regardless of method, you generally implement network-wide routing with the help of a carrier. Then you establish rules to select which sites calls go to; the term for this is "pre-call routing." You also set up rules, as you would in a single-site center, that determine which reps callers reach. The term for this is "post-call routing."

Your criteria for sending calls to each site can be static, if reps at certain sites answer calls at specific times of the day or week. You can also create dynamic rules that choose sites where reps are likely to be most readily available. A detailed discussion of network-wide is beyond the scope of this book, and we refer you to our list of recommended reading at the back of this book for further information about this topic.

Meshing Disparate Centers

If your company has multiple sites with phone switches and IVR systems from different manufacturers one option to consider, besides using third-party call routing software, is establishing a network of centers. There carrier-based routing makes sense. But if you prefer premises-based then what are your options?

You can rip out the old switches, which if they're 5 to 7 years old may make sense, but if they are newer you might want to keep them around because they are expensive to replace.

Or you can purchase middleware, such as Cisco's and Genesys Labs' products that are switch-independent. Middleware is an excellent option for premise-based routing if you are having calls escalated from outsourcers using different switching platforms. Of course if that outsourcer has built a dedicated center or chunk of one to you, you can specify the technology.

Teleworkers: Overcoming the Technology Hurdles

You're sold on teleworking because it saves you from providing your reps with costly and quickly marked-up coffee bars , otherwise known as workstations. Besides, if they're home they can't blame traffic for coming in late. Sorry, rushing out for that $100 bag of Bohemian blend coffee at the Seattle Drip, to recover from three hours' sleep because they watched the John Waters film festival on cable last night or tripping over the family furball stretched across the bathroom entrance doesn't count.

While the PC and Internet have made teleworking a viable option for support center reps, according to recent articles in *Call Center* magazine (which closely follows teleworking) there are still technology hurdles to surmount. Chief among these is the connection linking the teleworker with the support center. If we're talking PSTN, it can mean upwards of three lines, one for personal use and two for dedicated voice and data links to the support center. For reps working at, say, $12 per hour and having to foot the bill for equipment and business lines (not uncommon), the extra phone bills can be taxing.

What's more, PSTN lines may not be enough for high-volume applications support reps often need access to. High-speed connections to the home — principally digital subscriber line (DSL) and cable modems — satisfy the broadband requirement. But these can come at a hefty price tag, depending on the locale.

Adding to costs is the ancillary hardware and software that agents typically need to hitch onto the support center's voice and data networks. These encompass hardware adjuncts to reps' desk sets that connect to the corporate PBX and/or ACD, thereby enabling three-digit extension dialing, call transfer and park, conferencing, skills-based routing and other features. They also include software that reps download to PCs for data entry, PC call control and monitoring purposes; and, at the corporate end, a gateway or integrated access device (IAD) that transmits voice and data to the remote worker.

As Steve Schilling, ceo of application service provider Netifice (Norcross, GA) so aptly puts it: "The single most overlooked aspect of teleworking is providing technical support for remote workers. Stuff breaks down that you have to be able to fix."

Still telework solutions are maturing; they're becoming more reliable and delivering a growing range of office functionality. Apart from PBX/ACD access, many current products bring PC call control and/or CRM functionality to your reps' desktop. A good example is Rockwell Electronic Commerce's (Wood Dale, IL) LAN Agent, which offers a softphone to manage calls on-screen using a mouse.

Other examples could include:

Gematech's (San Diego, CA) Remote Service Manager (RSM) which routes phone calls to remote reps without a traditional ACD/PBX. You install the software on a server in your support center. When a customer calls your company,

RSM uses ANI information to identify the caller. The software then routes the phone call to the next available agent with the appropriate skills. RSM can also prioritize agents based on which agent the caller spoke with previously.

"A management system is very important in teleworking to allow supervisors to monitor home workers' activities," says Richard Floegel, CEO of GemaTech. RSM lets you access statistics like call queues and agents' statuses through a Web browser, network connection or dial-up. You can also create historical reports that provide information in graphs and charts.

To help monitor rep performance, GemaTech includes its Secure Voice Recording (SVR) software. SVR lets you monitor and record phone calls to listen to live or play back. Recordings are stored on the LAN.

RSM can handle up to 2,048 agents and between 15 and 120 simultaneous calls per server. For more capacity, you can link up to 255 servers while maintaining consolidated reporting. The product is also available as a hosted service.

MCK Communications' (Needham, MA) Mobile EXTender software, which lets users connect to the PBX from wireless or wireline touchtone phones. Using Mobile EXTender, agents can answer incoming calls, transfer calls, make conference calls and dial internal four- or five-digit extensions.

The software is compatible with Avaya, NEC, Nortel and Toshiba phone switches. To enable the software, you need an MCK PBX gateway, hardware that connects to your switch. Gateways are available in eight, 12 or 24 ports.

One MCK customer, eSupport Now (Charleston, MA), has four call centers with 100 reps providing customer support for companies in retail or financial services. eSupport Now considered allowing reps to work from home for several reasons.

"For a lot of people, telecommuting provides a lot more job satisfaction," says Chris Selland, eSupport Now's vice president of operations. Though Selland found MCK's products helpful in enabling telecommuting, he discovered that not all agents were suited for working from home.

"We still look at telecommuting positively," he says. "But the telecommuting workforce is fairly transient. [Telecommuting] requires a more dedicated agent and keeping these kinds of agents is costly."

Since eSupport Now had underused real estate, having agents work from home was not saving them office space expenses. However, eSupport is still using MCK for sales and executive employees, whose work is better suited for telecommuting, says Selland.

Linking Field Service with Customer Support

Field service has traditionally been closely linked to customer support centers. Today customers can try to resolve problems themselves and enlist the aid of a support center rep, but sometimes there is still a need for a friendly visit from a field service technician to solve their problem.

Your friendly neighborhood telco is the best example of this integration. Some problems can be solved by the customer, e.g. did you plug it in? Others are diagnosed by the support center (when they test the line), while other problems need to be checked out and fixed by the line personnel. Like too many people and too few lines serving your neighborhood, leading to overhead conversations on your supposedly private phone line.

Field service departments are increasingly dependent upon support centers to manage service calls, and many companies are taking the next logical step — automating the link between field service and support center systems.

Vendors know this and offer applications that provide some or all of the following types of functionality. These include installed product configuration management, service requests and work orders, service entitlements (service agreements, warrantees, etc), technician dispatch and scheduling,

spare parts inventory management and mobile & Web access.

With a virtual private network (VPN) at the helm, you could take this link a step further and provide a CRM-enable a field service link (see chapter 11 for more on CRM). That way everyone is literally on the same page. Much of today's 'CRM' software is the marriage of sales force automation with problem management/problem resolution software.

The Virtual Private Networks (VPN)

During the lifespan of this book the support center community will increasingly use VPNs as their link internally and externally. This means that the technical hassles and costs in linking support centers, whether with other support centers or with outsourcers and teleworkers, may finally fade away.

VPN technology provides the connectivity for remote access, intranets and extranets. It gives all types of communication — voice and data — a hackproof and high-speed channel. It can extend your center's network beyond the firewall to your reps' homes and provide secure, robust communications between your center(s) including all outsourced locations. It can give your support center a "private channel" on the Internet that can be shared by any authorized user because your telecommunications provider supplies your center with a protected network by technically extracting a part of the public network and labeling it for sole use of your support center's or (more likely) company's operations. This allows for better control routing of not only calls and data transfers but also telecom costs.

You may use a VPN and not even realize it. For example, CMP's laptop users (like co-author Brendan Read who works out of his home in Canada) enter their password to connect into the VPN to access the company's Lotus Notes program for their e-mail, the Internet and more. Employees like Read are truly virtual employees.

The VPN installed today will, in all likelihood be an 'IP VPN' — a virtual web of interconnected tunnels carrying encrypted traffic multipoint to mul-

A few vendors even offer linking capabilities whereby data generated in the field can find its way back into the support center's system to be analyzed and parsed.

Here's how that would work. Field service staff can feed information about work orders, service calls and customer reactions to new products back to CRM systems for analysis. CRM systems, provide the support center and the field service personnel with crucial information about dynamic events and customer

---

tipoint. This enables domesticated packet-switched voice to travel along with your center's text data, allowing your support centers, reps and tele-workers to be connected by a single set of pipes, without undue packet loss (the main cause of poor voice quality).

VPN technology provides the cheapest, fastest way to build readily remote-accessible voice/data WANs. Capable of yielding higher bandwidth connections than wireless / mobile connections, VPNs give reps working at home or at other support centers more of the performance and capabilities they'd enjoy if they were in the same location as the technology.

VPNs are not just for creating private data connections. VPN telephony, i.e. voice traffic with a high quality of service (QoS) can also traverse these private byways. (VPN telephony can run over IP, ATM or Frame Relay.) Using the corporate VPN to transmit all communications — data and voice — is a very appealing proposition, especially for the support center industry. The apparent cost savings that they can realize by combining voice and data onto one network is compelling.

Support centers are beginning to adopt VPNs. The vendors have taken note — there are products, such as for teleworking that are specifically designed to work with IP VPNs. MCK, Avaya (Basking Ridge, NJ) and Teltone (Bothell, WA) can facilitate corporate network access using an IP VPN connection.

Cisco Systems (San Jose, CA) understands the importance of VPN and VPN-based telephony in the call center marketplace. Cisco's IP call center software offers redundancy comparable to traditional call-routing systems. It's Intelligent Contact Management (ICM) software routes incoming calls to call center reps the same way a traditional automatic call distributor does, but offers cost, programmability and manageability advantages over a traditional ACD.

updates gathered from all touch points and departments. For example, a customer cancels a service call at the last minute, R&D discovers a possible bug in the equipment scheduled to be serviced or an easier way to repair a product is discovered by another field service representative.

Linking field service systems to a support center's systems also allow such personnel to play a more productive role. With the proper training, and incentives they can help up-sell (e.g. new equipment, service contracts) to customers, raising the customer relationship through field service operations to a new level.

Eventually, your support center's systems can be so tightly connected to field service systems that customers will go on-line and get a diagnosis to their problem, request a service visit appointment, a confirmation either immediately or by e-mail. Customers could also get an e-mail or phone call reminding them of the appointment 24 hours prior to the scheduled visit, or maybe even receive a notification when the field service person is en route.

When looking at field service components within a support center infrastructure, consider deploying work order dispatching, part order and reservations, preventative maintenance scheduling and real-time data transfer to and from wireless devices. Closely look at the product's ability to track inventory handled and assigned to field force staff, to access, record and track problem resolutions while in the field and to address device compatibility issues.

As the Canadian comedian Steve Smith, who plays the ultimate problem fixer Red Green on the *Red Green Show* (also carried by PBS) puts it: "We're all in this together."

Eight

CHAPTER EIGHT

STAFFING AND TRAINING

Staffing and Training

CHAPTER **Eight**

When you have knowledgeable support reps that stay with your company for the long term, you can translate your company's investment in a support center (real estate, furniture, equipment and software) into loyal customers.

Qualifications

The first step in building a staff of knowledgeable support reps is determining the precise qualifications you are looking for and matching them to actual job responsibilities.

These qualifications must accurately reflect the support needs of your customers. Rick Kilton, president of consulting firm RWK Enterprises (Lyons, CO), points out that at some support centers, the first support reps whom customers reach are often the least skilled. Even if reps are able to assist half of all customers immediately, that still means that the remaining customers are likely to end up speaking with several individuals before finding a rep who can help them.

Regardless of what products or services they support, the two skills all rep must have are the ability to solve problems and to serve customers.

Problem-solving skills require more than knowledge. When reps don't know the answers to customers' questions, they should know how to find them quickly.

Reps must also demonstrate empathy and patience when dealing with customers. Intentionally or not, reps can irritate customers when they give the impression they're trying to discover what your customers did wrong. The most effective support reps know how to listen and show they understand what customers are telling them.

"Some product geniuses may understand every bit of software code or every logic circuit in the hardware, but if they alienate the customer in the support process, they are a detriment to the mission of the support center," says RWK's Kilton.

These messages are getting across. According to the "2001 Best Practices Survey" published by the Help Desk Institute (HDI; Colorado Springs, CO), released in December 2001, soft skills such as listening, verbal, telephone customer service, questions, and problem solving skills were ranked as most important by over 90% of the respondents.

With more customers requesting support on-line, support reps should know how to communicate in writing accurately, politely and efficiently. At most firms where reps answer both calls and e-mail, they do so in separate shifts rather than having to respond to both at once.

As you establish your qualifications, take a careful look at the requirements for each level. Here are some general guidelines:

Level-Zero

You can define level-zero support in two ways. In one sense, level-zero support is automated assistance that lets customers find answers through an interactive voice response system or on-line knowledge base, alleviating the necessity of contacting a rep.

Or you can describe level-zero support as a job category that refers to the very first group of individuals whom customers reach. Level-zero reps are not support reps; their role is to discern why customers are contacting the company before directing them to your support organization. If your company charges for support, zero-level reps can also verify that customers are entitled to the help they're asking for from your support center.

Level-One

Level-one reps, by contrast, help customers with broad questions about a company's products and services. A level-one rep's temperament in working effectively with people and handling stressful situations is more important than specific knowledge of products.

Ideally, level-one reps should have at least two years of college and two years of work experience. They don't necessarily need college degrees. Some outsourcers like Stream International have been successful with training level-one reps who are high school apprentices and who had worked as coal miners.

Additionally, level-one reps need to have the discipline and tenacity to try to solve tough problems and work within a 24x7 environment that sometimes requires overtime.

"You typically don't offer this job to a person fresh out of high school," advises Mia Melanson, principal with Performance Consulting (Natick, MA). "You want someone with the maturity to understand how important providing a solution is to the customer and his or her business, and who can calmly handle irate customers."

Level-Two

Level-two reps must be able to analyze and diagnose difficult problems on their own rather than depending on a knowledge base. They should have ample experience, appropriate training, and the confidence to try and find answers beyond those you've already documented, including questions about problems that are unfamiliar or intermittent.

According to Melanson, they must know whom in the support organization or other business units to tap for help. In technical support centers, level-two reps usually have a strong computer science background. They have degrees from four-year colleges or Associates' degrees coupled with three or more years of technical support experience.

"Training depends upon the company's business, depth and complexity of product line, and specifically what the support organization is designed to do," she points out. "You may have a relatively flat organization with just two levels of support or you may have multiple levels including field support. And in some support organizations, level-two reps are also expected to coach level-one reps and be the role model for professionalism."

Level-Three

Level-three reps are subject matter experts; if they can't answer customers' questions, no one can.

Companies that provide technical support often look for level-three reps with computer science, electrical engineering or other four-year degrees. They also seek reps who know the intricacies of the products they support and understand how they work at the core level. They can typically write code or fix equipment.

"These reps are the 'black belts' of support," says Melanson. "Level three is the last stop before engineering and product development. In some organizations, these are the programmers. And they are expected to fix bugs or get to the root cause of a problem and do whatever it takes to eliminate it immediately or by the next release."

Wages

How you staff and train the customer support center staff has the biggest impact of any factor on the cost of your support operations. As we note elsewhere in this book, labor comprises from 70% upwards to 90% of a support center's costs.

To find and keep qualified support reps support centers typically pay wages that are as much as 15% higher than those at other types of call centers. Wages for support reps begin at $8 per hour or $16,640 per year in the US. In Canada, wages are $6 per hour or $12,480 per year.

For level-one support reps in major metropolitan areas, wages rise to between $13 and $15 per hour, or between $27,040 and $31,200 per year. Wages for level-two reps range from $20 to $25 per hour, or $41,600 to $52,000 per year.

Higher pay is commensurate with the amount of training a support job requires, but it also reflects competition for people with the same skills. In technical support, for example, competition comes from software developers, equipment makers and Internet service providers that can pay higher salaries than support centers.

Support center wages, like wages in other types of call centers, can fluctuate year to year. Figures vary depending on who conducts the research. Human resources consulting firm William M. Mercer (New York, NY) reports in its 2001 Call Center Compensation Survey that the average technical support wages of respondents dropped from $12.60 per hour in 2000 to $12.05 per hour in 2001.

The higher the level of the support rep you're trying to hire, the greater the skills the rep needs to succeed at these levels. It's in the best interest of support centers to keep reps as long as possible to avoid replacement costs. Turnover can cost support centers as much as 50% the cost of a rep's annual salary — in recruiting, screening, hiring and training a replacement.

Support centers have less annual turnover than other call centers. The 2001 William M. Mercer compensation report reveals that annual turnover is 51% in call centers that provide technical support, compared with 187% in centers that conduct outbound telemarketing, and 94% in centers that provide inbound and outbound customer service and sales. But, turnover among support reps is greater than it is in credit and collections centers, where the annual rate is 33%.

As with wages, annual turnover can vary from one year to the next. The 2000 William M. Mercer survey had reported a turnover rate of 30% for technical support, compared with 51% in credit and collections.

Recruiting

The more specialized the need, the more selective you have to be. Consequently, the largest pools are for the level-zero and level-one reps. Individuals from a wide variety of backgrounds and education can work successfully as level-one reps. They include ex-miners in Nova Scotia, Canada, to ex-Navy service personnel in Maine, to First Nations and high school students in British Columbia.

When examining labor pools, do not let the collars of applicants' shirts fool you. As a *New York Times* article on North Bend, OR-based CyberRep (then known as 800 Support) a few years ago revealed, the brightest people had gone into jobs in forestry, which had offered the highest wages.

Many smart people prefer to stay in communities with few opportunities because of family ties and for the quality of life. Victoria, British Columbia, Canada, where co-author Brendan Read has returned to live, is an excellent example.

Victoria and other communities on Vancouver Island have pleasant climates all year and offer laid-back lifestyles. Many people in British Columbia would never dream of moving to big-city Vancouver on the mainland. Read knows of several school friends from the University of Victoria, where he earned his degree, who work as cab drivers and store clerks because they want to stay in Victoria.

If you're trying to hire technical support reps, consider communities with military bases. Other sources of top-drawer talent for level-one or level-two

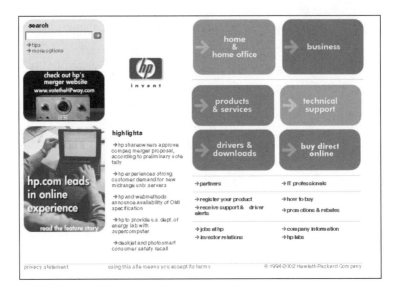

reps include outsourcers, which typically open centers in smaller communities with lower wages and less competition for labor.

"Experienced reps who've worked for service bureaus, like Stream International, have had to learn to fix a broad range of hardware and software," says Kilton. "They've shown they can easily learn any product."

To find support reps with the skills you need, you have to go where they hang out. These places include colleges, military bases and community organizations that assist job seekers. Look also at venues in cyberspace, whether they're general job sites like Monster.com or more specialized sites like CallCenterCareers.com.

Support centers, especially outsourcers, use job fairs to recruit and, more importantly, to test local labor markets. Human resources professionals caution, however, that while job fairs and newspaper classifieds work very well for customer service positions, they are not as effective as the other means we've mentioned for seeking support reps. And yet, you shouldn't overlook the traditional newspaper classifieds; job seekers can flip back or cut them out a lot more easily than Web pages. Remember, too, that people still listen to radio, and repeat ads in top markets stick in listeners' minds. It's also a good idea to offer open houses, either at your company or on-line, to get applicants through the physical door.

When recruiting, it helps if you have a name brand that people know and can identify, especially if you run a service bureau with a support contract, and the client allows you to publicize it as a reference. If your company doesn't have a well-known name, you will have to offer additional inducements, like better pay, amenities, facilities, work environments and career paths than your competition.

Once support reps begin to work for you, you can set into motion the most powerful recruitment tool there is: word of mouth. If your company has a great reputation, you'll have little trouble getting qualified reps; employees seek out and recommend potential applicants whom they think will work out. Employees also appreciate incentives like bonuses for those who recruit others who end up staying for at least six months or a year.

Staffing Agencies

Another recruiting option is to hire staffing agencies. They offer temporary positions, direct placement and insourcing services. They have pools of job seekers whom they have already recruited and screened, and they can recruit and screen new applicants rapidly. Companies often go to staffing agencies to find reps for level-one support, although these agencies can also find reps who can provide higher-level support.

Temping, also known as flexstaffing, entails contracting with the agencies that recruit, screen and assign support reps to work at your company.

Flexstaffing offers several key benefits. The most important benefit, as the name implies, is flexibility. This flexibility is essential if you have seasonal demand, like during the winter holidays.

Another benefit is that you directly supervise the reps without the hassle and headaches of recruiting and firing. You can bring in qualified reps when you need them, but you're not obligated to keep them on staff if they don't work out, or if your support needs diminish. Although you do have to provide the workstations, you maintain control over the equipment, data and communication with customers, which is not always the case when you outsource to service bureaus.

Going to staffing agencies is an excellent way to try out prospective employees. About 60% to 80% of the level-one support reps from Manpower go from temps to on-staff employees within six months or a year.

Insourcing, often used for internal support, is also practical for customer support. With insourcing, the staffing agency manages the entire support operation, but on your premises and under your direct supervision through the on-site manager.

Staffing agencies can help you ramp up new support centers. The good agencies know the best places to recruit locally, they can get the best people for your needs probably far faster than you can.

There are some downsides to staffing agencies. Support centers still have to spend money on training people who are not their own employees and who could leave at any time. You may not get the full payback on training temps as you might if they had worked for you directly.

There is sometimes friction between temps and staff, usually on the part of the temps who are resentful they're getting paid less than the staff. If that happens, Rick Kilton recommends that you contact the staffing agency immediately and ask that it engage in a discussion with the temps.

"Temps must understand that they elected to become a temp and they were told the ground rules," he explains. "They get paid less, but in return they have the freedom of entering into and out of temp contracts, without the necessity of giving notice. If they don't like a client, they can tell the agency that they want to be moved. Or they can quit. That's the price they pay for their flexibility."

To make the staffing agency relationship work you need to select these firms very carefully. You need to find out everything you can about their recruiting, screening and training methods, their size of pools and retention rates. You also need to get references from other similar support centers.

When you assess staffing agencies, confirm that their references come from the community in which the support center is or will be located. One staffing agency could be the place to work for in one community, but could be the bottom of the barrel in another.

Make sure the staffing agency you select has an excellent understanding of your own culture as well as skill requirements. See to it that it works closely with your human resources department.

Finally, you must train and treat temps the same as in-house reps doing the same work. "Oh, they're temps," is not an excuse to load them with the tough calls.

"No matter how you staff and what you pay them, if your employees and ours are doing the same jobs you must train and treat them equally," says Linda Lauritzen, director of global call center services marketing for staffing firm Manpower (Milwaukee, WI).

Training

To assist customers, reps receive training on every aspect of the product or service relevant to their level, and on the tools they will need to use. That especially applies when your company starts to support a new product or service.

"For example, a rep may have been able to perform at a high level in trou-

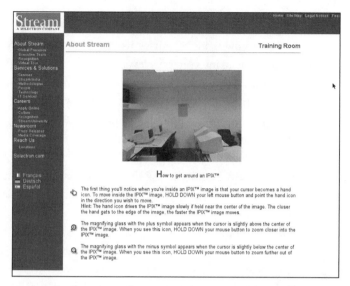

How to get around an IPIX™

The first thing you'll notice when you're inside an IPIX™ image is that your cursor becomes a hand icon. To move inside the IPIX™ image, HOLD DOWN your left mouse button and point the hand icon in the direction you wish to move.
Hint: The hand icon drives the IPIX™ image slowly if held near the center of the image. The closer the hand gets to the edge of the image, the faster the IPIX™ image moves.

The magnifying glass with the plus symbol appears when the cursor is slightly above the center of the IPIX™ image. When you see this icon, HOLD DOWN your mouse button to zoom closer into the IPIX™ image.

The magnifying glass with the minus symbol appears when the cursor is slightly below the center of the IPIX™ image. When you see this icon, HOLD DOWN your mouse button to zoom further out of the IPIX™ image.

No matter how brilliant support reps are, they require training in dedicated rooms, like those at Stream's support center in Memphis, TN. Even a brilliant level-three rep needs to develop skills in customer service, plus refresher training and some time to get up to speed.

bleshooting product A, but will have to learn product B before applying those troubleshooting skills," explains Kilton. "Therefore, the training required for each level is determined by the skills required for that particular level or functionality of service."

Make sure you budget enough resources for training — both new-hires and ongoing. Training consultants such as Elizabeth Ahearn, president of The Radclyffe Group (Fairfield, NJ), say that it costs $2,000 to train each new hire and between $800 and $1,000 to provide ongoing refresher training per agent per year to enable agents to keep up their skills.

Here are some general guidelines for training by level:

Level-One

The primary job of level-one reps is to find out what information customers are looking for. They don't have to know the answers to all questions or solutions to all problems.

At the very least, level-one reps should know how to search through an on-line knowledge base, if your company has one, to determine if a customer's question has a straightforward answer.

Training depends on how many inquiries from customers you want level-one reps to answer and how many you want level-two reps to handle, points out Rick Kilton. If you want to avoid escalations, then you have to seek out better-qualified level-one reps, train level-one reps on a broader selection of products or further develop reps' expertise on specific products.

There's a price for that: more level-one qualifications can limit your labor pool and require that you pay higher wages.

"Level-one reps require the ability to ask the right questions and induce the right answers," explains Fred Van Bennekom, principal with Great Brook Consulting (Bolton, MA).

If customers can view a portion of your knowledge base from your Web site, and if you require customers to identify themselves before they do so, check whether your knowledge base software lets reps see what areas of your Web site customers have looked up. This way, reps don't inadvertently offer information that customers have already consulted but may not have found helpful.

Since they are often the new hires in a support center, level-one reps usually have the least training and experience. After between six months and a year, they can earn the opportunity to specialize, which frequently means a promotion to a higher level.

Level-one reps generally learn from colleagues, supervisors and information they gather from the company's knowledge base. Many organizations that provide support for tangible items like clothes or stereos co-locate their warehouses with their support centers so that they can train reps quickly on their products. If the items are small enough, companies keep samples of products within easy reach of reps so that they literally have a grasp on customers' questions.

It is usually less important for level-one reps to understand how a product or service works. "Most level-one calls are simple and often involve installation or use of products or services," explains Geri Gantman, a senior consultant with Oetting and Company (New York, NY).

Level-Two

Level-two reps need training on how and why products and services work, as well as the typical questions asked and problems customers are likely to experience.

Mia Melanson says it's best when level-two reps earn promotions from level-one positions or from elsewhere in the company since they already have experience with the company's products and communicating with its customers.

For level-two reps who rise from the ranks within your support center, refresher training is helpful, particularly when it emphasizes how to handle challenging questions from customers who, after spending a lot of time speaking with other reps, may be close to losing their patience.

Level-two reps should also receive extensive training on all new products, how they are supposed to work and what really happens when customers use them. Lastly, they need to describe customers' questions and their respective answers in writing both accurately and thoroughly. This documentation is essential, as it serves as the raw material for your company's knowledge base.

In companies that expect level-two reps to take on leadership roles within the support team, reps also need to communicate with diplomacy and be able to teach or coach others. Reps can develop these skills on the job, and through a combination of workshops and one-on-one meetings with managers.

You may also want to train level-two reps on niche products and services, Kilton suggests. This way, level-one reps can answer the majority of customers' questions, and level-two reps can deal with the relatively smaller percentage of questions that require expertise.

"With niche products, you will need level-two reps who can really get into the product and know how it works because the level-one reps will not have enough day-to-day familiarity with how to fix them," Kilton says.

Level-Three

For level-three reps, training is truly on the job, says Melanson. They already have the knowledge and over time they gain troubleshooting skills with each support request their colleagues escalate to them.

Technical support centers employ level-three reps to debug software, fix products, and where necessary, write software to work around problems. An important, but often lacking, skill is the ability to work as a team with other levels of support.

Kilton suggests that you deploy level-three reps on versions of products that are no longer for sale but for which there is still an installed base. Lower-level reps are more appropriate for fielding the larger number of questions about products that are still on the market. There is no point in training lower-level reps on outdated products with a dwindling base of customers.

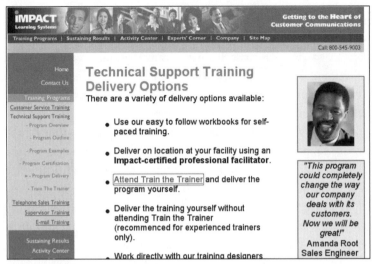

Impact Learning Systems (San Luis Obispo, CA) offers several different delivery choices for its support curriculum: workbooks, on-site training, train-the-trainer program and customized on-line courses.

Training Tools

To train reps, support centers combine classroom courses, off-site seminars, at-seat coaching, together known as facilitator-led training (FLT) and technology-based training (TBT) tools. They include videos, CDs, on-line courses that simulate communication with customers and Webinars that allow reps to connect with an instructor in a group setting.

Prepackaged courses, formerly available only on CD-ROMs, are now becoming available on-line. Some developers host their training software from their servers to make it easier for support centers to administer and customize their courses.

Technology-based training from reps' computers is effective at spreading knowledge, such as on how to use a product. It is less costly and less disruptive to reps' schedules than classes with live instructors. Reps learn at their own pace at their own desks, including, at times, when they are not receiving lots of support requests from customers.

Technology-based training is convenient, Melanson says, because it does not require reps to be out of the office for several days to attend off-site workshops.

"And since customer support reps enjoy tinkering with technology, it is a natural delivery mechanism," she says. "They tend to get a lot out of technology-based training tools."

Customer support centers do not differ much in general from other types of call centers in the methods and tools they use to screen and train reps. They rely on assessments and simulations to select applicants. Mia Melanson recommends simulations and interview questions that test analytical and problem-solving skills.

TBT proponents claim that these technologies improve information retention, reduce errors, minimize escalations and cut staff turnover. Reps can refresh themselves with courses on their computers, which is far easier than remembering what the instructor taught or rewinding a tape.

If you allow reps to work from their homes, on-line courses are the most practical options unless you require reps to receive training at your support center.

Anne Nickerson, who heads up Call Center Coach (Ellington, CT), a training firm, recommends that call centers ask themselves the following when they consider training software: Does the tool model behaviors you'd like to encourage among reps? Does the vendor offer implementation and post-implementation services? Are the skills and behaviors measurable, or at least observable? And is the product flexible enough to incorporate your support center's unique goals?

To avoid inconsistencies in the quality of training, Nickerson advises setting guidelines for the content that supervisors choose. She recommends that support centers set up a training implementation team of supervisors and coaches to review and approve content.

"The role of supervisors is to supervise; they're not professional trainers," she says. They should only use packages whose training has been validated by outside experience."

Call Center Coach's Nickerson says call centers with 50 seats or fewer may find the products too costly, unless the vendor can host the software on a monthly basis.

"Senior training managers [who] think they can replace trainers with computer learning are making a grave mistake," she says. "Rather, trainers need retooling to learn how to manage group dynamics and facilitated discussions. They also need to know how to differentiate the quality of the multitude of training methodologies and products."

Rick Kilton says FLT instruction is better for developing skills. Students in classes learn from each other, and the instructor observes how reps learn most effectively. Reps can get answers to questions right away, and role-playing is an option.

With live instructors, students can't cheat or anticipate answers to multiple-choice questions as easily as they can with automated courses. A good instructor pays attention to how reps behave, not just what they say.

"As an instructor, I prefer to see the students to see their expressions and other body language," says Kilton. "If I am administering a virtual classroom, how do I know a student isn't answering e-mail during the class? How do I know if a student is puzzled but afraid to raise a hand? How do I know if a student is not interested?"

Many developers of training software don't recommend completely replacing FLT with software. They agree that a blend of live and automated training works best. Reps can receive personal attention from a classroom instructor, and to speed the learning curve, reps can use training software to review what they've studied. Vendors also say that blended training is a common practice among clients.

There are many facets of customer support that can't be taught on a machine, such as those that involve customer and employee management. The Service and Support Professional Association Certified Support Specialist training is FLT-only.

"I don't think that classroom training will go away," Frank Russell, president and chief executive officer of GeoLearning (West Des Moines, IA), told *Call Center* magazine associate editor Lee Hollman, in an article on computer-based training that appeared in the March 2001 issue. "I do think we're moving toward a blended world, where part of training can be done on-line and part of it can be done live."

Russell recommends software to achieve consistency in training. Some classroom instructors are excellent, but most are likely to be mediocre, if not worse. Although courses on computers aren't as intimate as classroom instruction, they offer an interactive way for reps to discover how their behavior affects customers.

"You don't have the ability to read faces, so you need to find ways to [get reps to] participate and learn by interacting," adds John Walber, chief operating office of HorizonLive (New York, NY). "You need to ask more questions, throw out some open-ended concepts [and let] students chat on their own."

Communications Training

All reps have to learn how to use your center's phone system, software packages and knowledge bases. They should also receive training on how to use the Internet to find information and solve problems.

On-site training that helps reps learn about themselves and how they relate to others.

Given the increasing use of e-mail and chat in support centers, reps also need to know how to communicate in writing. You should screen all reps for grammar and spelling skills, and recommend remedial training for those who fail to make the grade.

Ideally, reps should have a minimum of an eleventh-grade reading level. The higher-level the rep, the higher the reading and comprehension levels reps must have.

Train all your reps on how to use e-mail correctly. That means avoiding slang, jargon and condescension, not to mention ALL CAPS. Language must be formal, polite and solicitous. Reps neither see customers, nor do they hear inflections in their voices, so they can't assume customers' on-line correspondence represents all they're trying to communicate. It is nearly impossible to "read" an on-line customer.

A key part of e-mail training, say consultants, is teaching when the reps should pick up the phone or ask customers to call them. Reps must attune themselves to customers' choices of words, like technical terminology or phrasing that suggests frustration.

"The reps must sense that frustration early on and contact the customer directly before he or she walks away, or escalates the situation," says Mia Melanson.

Training Reps To Be Human

If your company's focus is on technical support, you may need to teach techies to be human, or pass as humans.

Rick Kilton advises companies to train reps on how to get customers to cooperate with them so they can understand the customers' problems; this includes listening for vital information. Then the reps need to learn to judge how long they should devote to assisting each customer. These methods can avoid multiple requests from the same customer about the same issue, or unnecessary escalations.

He also advises support centers to teach reps to verify with the customers what the reps thought the customers said to ensure both understand one other.

"Provide the time and attention it takes to help your customer with their problem or question," he advises. "Any time a call is rushed, it will be incomplete — often requiring [further communication] from or to your client. Sometimes, only another five or ten minutes will produce the solution and will not only prevent another call but will create a satisfied customer."

Also, train reps to acknowledge when they are unsure of an answer to a customer's questions. What customers care most about is that the support rep is doing his or her best to find answer to their problem; they don't expect reps to know everything.

"Sometimes, more work is needed on the part of your company to research or replicate the problem," explains Kilton." Explain why they must perform some more steps to help isolate the problem. Explain why that will help both of

The Customer Support Clash

There is often a clash between problem-solving and customer service skills. Training experts say that it's more notable in technical support, knowledge does not necessarily translate into communication.

"Unfortunately, what you have is a clash," says Rosanne D'Ausilio, president and founder of training firm Human Technologies Global (Carmel, NY), "you have customers who are very stressed out and defensive because they have problems they can't fix. And you have reps who are very knowledgeable and can solve the problems, but who come across as arrogant and condescending — just what the customers don't need."

Reps may have the answers, but customers may not listen. The results: the problems remain unsolved. If customers are mad enough, they'll go to competitors and badmouth the companies that employed reps who they perceived were impatient or condescending.

"Customers want to be treated with dignity and respect," says Dianne Durkin, president and founder of training firm Loyalty Factor (New Castle, NH). "They want to be assured that the person clearly understands their problem and will be able to provide them with a solution. They want to feel understood."

Rick Kilton, who runs a consultancy that provides training for support centers, observes that more companies seek applicants with customer service and people skills, not just product knowledge.

"Companies are now figuring out that when their reps deal with customers with respect, the customers become cooperative, which helps solve the problems," he says. "Even if reps don't solve that problem, if they are respectful and friendly, that customer is more likely to stay loyal than if the reps are cold and hostile [but] fix the customer's problem."

you find a solution. Involve the customers in the decision-making and they will feel ownership of the problem and cooperate."

Reps should expect that customers don't always understand the products they operate, and they should be careful to not let their expertise go to their heads. Instead, reps should take the time to understand customers' knowledge of the products they use.

"The customers are likely experts in their own fields, like finance, accounting, sales, manufacturing and distribution," Kilton points out. "If they were experts on the technical products, the reps wouldn't be needed. No one appreciates another person being condescending. It is one of the most destructive acts a support person can do. Remember that everyone deserves to be treated with courtesy and respect."

There are many training programs that enable you to develop support reps' customer service skills. A program that training firm Loyalty Factor (New Castle, NH) developed and implemented has helped the company's clients improve customer satisfaction and service levels and lower turnover rates. In one instance, notes the firm's president, Dianne Durkin, Loyalty Factor reduced a client's annual turnover from 65% to 35%.

This company's program offers four weekly sessions of on-site training that helps reps learn about themselves and how they relate to others. Training introduces reps to various styles of communication styles, helps reps figure out what styles they use and discover which styles complement theirs.

Numerous exercises like role-playing give reps the opportunity to put themselves in their customers' shoes. One exercise in particular illustrates the importance of asking customers questions.

During the exercise, reps go to the front of the class and describe images for their colleagues to draw. In the first stage, reps don't ask any questions, so they must rely entirely on the rep's description to complete the drawing. The resulting images, says Durkin, are quite different from the original images. Only when reps ask questions do their images begin to resemble the initial drawings.

"We teach reps about different ways that people process information, and encourage them to listen for clues when the customer speaks," Durkin explains. "Then they can respond to the customer in a way that will make them feel comfortable and understood."

Adds Durkin: "For example, if a customer uses visual words like 'look' or 'clearly,' the rep knows to describe the problem in visual terms. We teach them to stay tuned in for word choices that are auditory or emotional in nature, and then to respond by using similar phrases."

Such soft skills training pays off by slicing talk time and avoiding escalations. Using Rosanne D'Ausilio's doctorate thesis — "The Impact of Conflict Management Training on Customer Service Delivery" at California Coast University in 1996 (based on providing such training for a utility) as a guide, the company saved over $330,000 annually by slicing talk time by 22.3 seconds.

She taught agents handling customer complaints conflict management strategies that enabled them to resolve the issue on the first call. "The company could handle more calls with the same number of employees," she explains. "The 22.3 seconds shaved off each call annually is equivalent to having seven extra full time agents according to the utility's calculations."

Certification

To determine if an applicant is suitable for the job, and to give reps career goals in the hopes of retaining them, an increasingly popular option is to certify reps and supervisors, and seek certified reps and supervisors when hiring.

Certifications based on objective standards can reduce the amount of time support centers spend on recruiting. Certificates give reps and supervisors badges to wear with pride and credentials to include in job applications and resumes.

There are two different types of certification, *professional* and *product*.

Traditional certification, specifically for technical support centers, internal or external, emphasizes knowledge of products, whether from IBM, Microsoft or Sun Microsystems. To earn certificates, technical support reps have to pass tests, usually with the help of courses.

Product certification is more relevant for internal support than customer support organizations. That's because internal organizations must support the products their companies buy.

Product training and certification can save customers' and reps' time if the training enables the reps to solve all of the problems immediately instead of asking customers to contact vendors themselves. Customers prefer one-stop support.

"For vendors such as Microsoft and others, certification is a good idea because it reduces the number of contacts with their support centers, and increases the knowledge of the customer in terms of how the products are intended to be used," says Melanson.

Take on-line services, for instance. In the mid- and late 1990s, if equipment at a local phone company's central office was the cause of problems with customers' Internet service, the provider of the Internet service would hand off support to the phone company. That's not as likely now, given that more phone and cable companies provide Internet service.

Many IT professionals are also skeptical of certification. "Just because you pass a test doesn't mean you know anything" is the common refrain, sung for good reason.

For example, co-author Joe Fleischer has a certificate in computer programming from Columbia University.

"Does that mean that I can get a job programming?" he asks. "The answer is probably yes. Does that make me a programmer? No, it doesn't. The reason is that I don't program; I just have a piece of paper that *says* I can do it."

Professional certifications evaluate broader skills. For support reps, they involve how to communicate with customers. For supervisors and managers, these skills include coaching and monitoring reps, setting up call routing schemes, scheduling reps, planning strategically and building relationships with other departments in the company.

According to Bill Rose, founder and executive director of Service and Support Professionals Association (SSPA; San Diego, CA), this can be a tall order for some people, and is why training and certification are necessary.

"Often there are support help reps who have excellent technical skills but do not have good people or corporate skills," explains Rose. "They can get impatient with customers who are usually not as knowledgeable as them. Sometimes they go out of bounds in their opinions, like saying there are flaws with the product. That may be inaccurate and does not solve the problem at hand."

The SSPA, The Help Desk Institute and STI Knowledge (Atlanta, GA) are among the leading certifiers of support. These organizations mostly focus on certifying reps and managers who provide technical support. One exception is the SSPA's certification program, *Customer Service Qualified (CSQ)*, which tests level-zero reps on basic service skills such as directing and escalating calls to support reps.

Each certification program tests for different skills. For example, organizations evaluate level-two reps and support center managers on strategic planning. The HDI's level-two program, *Help Desk Senior Analyst*, asks reps to design a new support center, analyze an existing center and consult on performance enhancements. The SSPA's test, the *Certified Support Specialist*, requires candidates to show teamwork and mentoring skills.

Many certifying firms also test managers and executives on skills like service level management, team building, budgeting, staffing and planning.

Certifiers usually offer a choice of testing sites: at your company, on-line or at off-site locations like schools or conferences. To keep their standards up-to-date, certification organizations often require support reps to take exams periodically to maintain their credentials.

Many certifiers offer TBT/online and instructor-led training to accompany certification. For example, the SSPA charges $150.00 for CSQ online training, plus $99.00 for testing and $299.00 for single (non-corporate) users to take the Certified Support Professional training, plus the test. For the more advanced Customer Support Specialist certification, the SSPA offers its instructor-led only course, for $8,900 for 20 students.

Keep in mind that definitions of and criteria for certificates can change. For example, HDI changed the name of its level-two certification program in 2002 from Help Desk Support Engineer to Help Desk Senior Analyst. It also rewrote the standards. HDI said its organization, like Microsoft and others, found that firms in some countries, most notably Canada, rejected the title "engineer."

Selecting Certifications

But with so many certifications, which one is right for your center? Tough question.

The question applies more to professional certification than to product certification. If reps need to know how to use a product to help customers with it, then they benefit from certification on that product, if it's available.

The problem with the current array of competing certificate programs is that managers do not really know or care which credentials are meaningful. It is almost no better than not having any form of certification. Support center managers and human resources personnel have to evaluate each program separately, and the inevitable question is: why bother?

"How does any manager know which one is best or appropriate since they are not likely to test them all?" asks Rick Kilton.

Kilton and others have called for a single industry standard for professional certification, much like those for certified financial planners or accountants. But that standard is unlikely to emerge any time soon.

In the meantime, when evaluating certifications, Kilton says you must consider that certification is in part a marketing program to generate other business, like training, which many certifying firms offer. Organizations that confer professional certificates want to make it possible for reps to earn them.

"Being an instructor for a large part of my life, I know that instructors teach the test, consciously or un-consciously," he says.

Another question to ask is how much do your customers value professional certification. They may have heard of Microsoft or Sun product certifications and regard them with some respect because IT professionals take these tests and because, in the words of one longtime pro, "these companies do a great job of marketing the certifications." But that's it.

"I suspect that most of them do not have a clue that professional support desk certification even exists," Kilton points out. "What they do remember is how they were treated."

Mia Melanson finds that professional certification is a valuable tactic for setting standards in support, as well as for encouraging the kind of recognition that attracts and retains new people who join the field.

"But professional certification is not an accurate screening tool because there are so many other variables like company or product knowledge, and on-the-job performance, which are difficult to predict with a multiple-choice test," she points out.

She notes that various companies offer certification for different purposes, to improve technical expertise or increase professionalism.

"It's important that support professionals recognize the realistic benefits of various certification programs and the intended purpose of each," cautions Melanson.

Retention

Because customer support can be an interesting and challenging job, and pays better than other call center jobs, turnover is less. But when turnover occurs, it costs more.

One reason turnover occurs, according to Melanson, is that customer support is still a new profession. School-age kids don't plan to be customer support reps or managers when they grow up. Most customer support professionals today did not necessarily select the profession among their first choices of careers.

"People become customer support professionals because it is a viable profession that combines people contact with technology," says Melanson. "Many customer support professionals get a kick out of solving problems and helping people at the same time. And customer support can lead to work in product development or other more traditional careers within companies."

To allow reps to get up to speed and apply their training, companies need to retain them for at least 12 to 18 months. To do so, they need to build career paths.

Supervisors also need to identify reps with strong skills in managing people and encourage them to become coaches and trainers.

Rick Kilton recommends setting up pay grades tied to performance, training and experience, with a minimum, like in the civil service. If, for example, a company offers premium support, like extensions to warranties or 24x7 assistance to customers who are willing to pay for it, then reps who help these valued customers have opportunities to advance to account managers.

"People want to feel that they are making progress," he says.

Companies also need to see interest and potential in management and product development. Many companies offer tuition reimbursement, but few employees take them up on it because long work hours makes it difficult to attend these classes.

Melanson contends that companies profit when programmers and engineers initially work in customer support because they learn firsthand how customers use products and about the problems customers experience with products. "If they had support experience and understand issues from the customers' experience, then they could gain the mindset and the insight to produce better products that also reduce support costs," she says.

Career Pathing

Career pathing is perhaps the best rep attraction and retention method of them all. Technical support reps who know how to fix products and know how they work can offer valuable insight into how to make them better.

But there are risks with offering career pathing. If you make promotions too easy, you risk greater turnover and understaffed support.

Kilton recommends that you set expectations from the beginning and explain the career paths in the company.

Make sure support reps first prove themselves in their jobs. Too many reps, especially with degrees, want a stepping-stone into the company, and may bail out as soon as the job they want opens up.

"I have advised clients to tell the new applicant that the support position is at least three or four years' minimum," he says. "Then see if they still want the job."

Advancing Quality Through Teleworking

A good way to attract and retain reps is to allow them to work from their homes. We discuss teleworking throughout this book, and it's worth mentioning here.

Teleworking eliminates costly and stressful commuting. It substitutes

Gift Programs

Rewarding reps and teams with gifts is a popular method to encourage excellent performance and retention.

According to the Incoming Calls Management Institute (Annapolis, MD), gifts rank third in the incentives offered by call centers — behind recognition and awards, but ahead of pay increases and job promotion.

How to supply gifts to deserving reps and teams? You can purchase the items, in consultation with team leaders, and store and distribute the goods. Or you can outsource the gift program to incentives companies that partner with product or service vendors and handle fulfillment.

Here's how they usually work: You assign values (e.g., dollars or points based on the performance you are measuring). Then you distribute awards directly to employees through certificates or tokens; or indirectly via the incentives firm that accumulates the awards in on-line accounts.

Employees can redeem awards directly through gift certificates or indirectly by, for example, purchasing items in program-specific catalogs with the points collected.

Gift certificates work especially well when they are for well-known firms with a broad range of products that appeal to a wide variety of employees. Because your agents redeem certificates on the spot, there is no waiting for redemption.

For example, TJX Corporate Incentives (Framingham, MA) administers improved quality of life for career advancement as a goal and performance inducement.

Anyone whose commute is more than 45 minutes each way knows how much life it saps, and how costly travel to and from work can be. As small cities become strangled in sprawl, and as high bandwidth connections become more widely available, teleworking becomes a more viable alternative to being at the office.

Teleworking also permits companies to retain reps when life situations change, like when a rep becomes pregnant or must take care of an ailing family member.

---

gift certificate programs for TJ Maxx, Marshalls and HomeGoods retailers.

These stores sell brand-name family apparel, accessories, domestics, housewares, shoes, giftware and jewelry, priced at 20% to 60% below prices in department and specialty stores. TJX-owned retailers number some 1,300 locations nationwide.

The Bill Sims Company (Irmo, SC) champions gifts because, unlike cash and gift certificates, they are tax-free to employers and employees, with some restrictions. The firm offers a quality gift catalog to clients.

"Employees also tend to remember gifts more than cash or gift certificates," says President Bill Sims Jr. "Also my experience has been that 15% to 20% of all gift certificates are never redeemed."

Some incentives firms go beyond point-gathering and gift-distributing. Through experienced call center industry partners that assess hiring and performance, RYI Solutions (Irving, TX) analyzes your staffing, training and retention issues. Once problems in these areas have been identified and corrected, RYI Solutions helps you devise and implement a gifts program tailored for your support center.

It offers a Web-based hosted solution that taps into your workforce management program to automatically track performance and convert it to points. An on-line catalog, browsed by employees off-line (say, in the break room) enables you to customize gifts to your center by gathering clicks and page views.

"Retaining employees and improving performance is not about how many DVDs you award," President Bill Campbell III says. "That's why we take a holistic approach. We want to help companies find out why employees are leaving and what can they do about it."

Support reps are generally suitable for and receptive to teleworking. The best support reps can work alone without supervision and like to be in control of their own surroundings.

Mia Melanson points out that teleworking is not effective in all settings. Reps who lack experience or need in-person coaching, live training and collegial support are not good candidates for a telework program.

"In a support center, teamwork really does increase productivity because team members can easily talk to each other about customers' problems and help one another solve them," she says. "If team members are working from home or remote sites, it becomes much more difficult."

Proponents of teleworking say it can improve productivity and dramatically lower turnover. According to teleworking consultant Jack Heacock (Parker, CO), some companies have found that productivity at their support centers increased by 40% to 80% when reps work from their homes.

But Melanson believes teleworking has limited use in support centers.

"There is a vital synergy by people working together in a support center," she says. "They can go to each other with problems much more easily than if they worked at home."

One aspect to think about is that reps need to have samples on hand. Is that option feasible for your teleworkers? If the support center fixes hardware and part of the process is to have equipment samples on the floor, it makes more sense to have the gear in one spot rather than in every rep's home. But if the gear is small consumer electronics items like portable CD players, then teleworkers can easily replicate the conditions customers are describing to them.

Ultimately, the decision to telework comes down to individual preferences. Consider: co-authors of this book, Joe Fleischer and Brendan Read, have access to the same phone and e-mail systems. Their employer, CMP Media, gives them the choice of working from their homes.

Joe likes being in a physical office, around people, but he occasionally works from home when the situation requires it. Brendan likes his co-workers too, but he enjoys what he says is a greater quality of life by working from home.

Nine

CHAPTER NINE

CERTIFICATION AND
BENCHMARKING

Certification and Benchmarking

CHAPTER **Nine**

In professional sports, athletes have to know where they stand. For example, where players shoot from on a basketball court determines if they score two points or three points. In a close game, that's the difference between winning and losing.

But what if basketball courts didn't have markings to enable players to distinguish between the two- or three-point range? Players would continue to maintain the goal of winning, but they would lose the ability to judge how they should move around the court. And no one would know how to keep score.

Now imagine that professional basketball teams never had to play against each other. Instead, teams could have as many players as they could fit on their courts and within their budgets. Teams would provide players with their own basketballs to shoot for the duration of each game. The players would earn rewards for scoring the most baskets in the least amount of time.

Unfortunately for would-be fans, the games would not lead to a championship, as these teams, citing competitive reasons, would decline to reveal the number of baskets they scored.

Sounds similar to how support centers sometimes measure themselves, doesn't it? Granted, customer service is not a spectator sport, but support centers benefit when they share information with one another, or at least adhere to a common set of standards. Hence the value of benchmarking and certification.

Organizations that offer benchmarking and certification comprise outside experts who take an objective look at your support operation and identify opportunities for improvement.

Because certification standards often emerge as the result of benchmarking, let's first establish how support centers benchmark.

What is Benchmarking?

Let's get one thing straight: You can't benchmark against yourself. You can compare data for your support center between the past and present. But to benchmark, you have to compare your center with others that have something in common with it.

Benchmarking firms can help. They invite companies to participate in their research, which includes assisting these companies' support centers with gathering data such as staff turnover rates, call lengths and service levels.

After collecting information about a sufficient number and variety of centers, the benchmarking firms segment these centers based on characteristics like size, industry, reasons for calls or types of customers.

Benchmarking firms also share best practices among centers that consistently achieve exemplary results, like the shortest hold times or highest percentages of calls that become sales.

Think your center is unique and doesn't fit into any categories? Don't rule out benchmarking. You'll be hard-pressed to find better information about best practices in support centers. Plus, benchmarking gives you a perspective that originates from outside your company, yet has special relevance to your center.

Something to Shoot For

The well-worn phrase in benchmarking is that you can only manage what you can measure.

Whether you run a support center for a company that fixes copiers or sells health care products, you're not likely to convince anyone you're doing a good job if you state that goals were met or customer satisfaction was increased.

With benchmarking, you first have to define outcomes that your support center can achieve through specific actions. For example, your center can't control the weather, but it can set up a routing algorithm to ensure calls arrive at alternate locations if the weather prevents callers from reaching reps at your site.

You should also be able to observe the outcomes you aim for. Every company wants happier customers, but happiness is difficult to evaluate. Many companies send satisfaction surveys to customers, hoping the feedback yields clues about customers' willingness to do business with them. But what customers indicate in a survey doesn't guarantee what they'll do in the future; it reveals how they recall their experiences with your company in the past.

Nevertheless, asking customers what they consider important, and getting their suggestions on where your company can improve, does help you identify goals you may not have recognized as priorities.

Surveys work best if they ask representative groups of customers a mix of broad questions, like how they perceive your service, and detailed questions, like how satisfied they were with the time it took to speak with reps who could assist them.

To be able to benchmark data, whether it's hold times or responses to survey questions, you have to produce outcomes you can evaluate. Qualitative responses are helpful, especially if customers offer recommendations your center can act on. But to have any value for benchmarking, the information you collect must give you a point of comparison.

Context Matters

After you identify outcomes you can evaluate over time, the next step is to make sense of the data you're gathering.

Let's start off this process by posing the question: Can you earn a promotion by saying your center improved first-call resolution by 100%?

Maybe, but this piece of data alone isn't enough to earn you the big bucks.

If they don't already know, executives want to find out the center's previous first-call resolution rate, and how long it took for the improvement to take place. They will also wonder how your center's resolution rates relate to those of other centers, and what basis there is for the comparison.

Most importantly, they want to know exactly what you mean by first-call resolution so that they understand what your center had to do to improve it. For example, customer support centers often define a first-call resolution rate as the percentage of requests where one support rep, after one communication with a customer, indicates that he or she fully accomplished what the customer asked.

To track first-call resolution, support center managers depend on information from support reps, who usually use software to create case files that correspond with customers' requests. If reps perceive they have fulfilled a customer's request, they can close the case, and if that closure occurs after only one conversation, then that's a documented, measurable instance of first-call resolution.

Or is it? Unlike static metrics like the number of calls in a previous month, first-call resolution rates can change at any time. If a customer calls the support center after a rep closes a case, the rep has to reopen it. If the rep already categorized the case as resolved on the first try, that designation no longer applies.

Another issue with first-call resolution is that in some instances, this metric tells you more about the quality of your company's product or service than your center's efficiency. If your company introduces new products, and you don't yet know what questions customers are likely to ask about them, you may need to look at first-call resolution rates over weeks or months to determine your center's effect on this metric.

Let's reformulate our question from the beginning of this section: How do you make an increase to your first-call resolution rate meaningful? Answer: Fill in the blanks by describing your efforts to bring about this change.

Your colleagues learn a lot, for example, if you mention you evaluate reps on their knowledge of particular products. With this information, colleagues appreciate why you changed the center's routing rules so that instead of sending callers to the first available reps, your routing software selects reps who rank highest on handling calls about certain products.

But be careful how you relate cause and effect. Circumstances outside your control may affect the outcomes of actions you've taken based on benchmarking. If first-call resolution rates go up during a month when your company discontinues all but one product, don't be quick to give all the credit to your new call routing algorithm.

When you compare present data with past data, verify all your definitions are consistent during the period you're looking at.

Let's say that the percentage of calls your center answers currently refers to the number of calls from customers who reach reps. If at the start of your comparison, the number of calls included calls from employees, but subsequent data does not, then you're contrasting different types of metrics.

Lastly, remember that blanket metrics are worth communicating to senior managers, who don't always have time to review details. But they should not be the basis for making decisions, which require more information, not less.

Check if the metrics you're studying vary with different types of calls. It's possible, for instance, that customers wait longer to speak to reps during some calls than during others. If you're not sure these discrepancies are good for your business, find out the level of specificity with which benchmarking organizations compare metrics. You could discover opportunities for improvement that appear small but turn out to have greater impact than you expect.

Duplicating Best Practices

The greatest gain from benchmarking is what you learn from your counterparts in other support centers.

Xerox offers a prime example, as it has been benchmarking its products, as well as other areas of the company, for several decades. In 1989 and 1997, the company received the top competitive prize the US government gives to businesses, the Malcolm Baldrige award, named for the late secretary of commerce under President Reagan.

Xerox's benchmarking efforts within its support operation began four years ago. The company has three sites, known as "welcome centers," that are the first places business and consumer customers reach before they receive support on copiers, printers and software associated with these products.

The centers are in Dallas, TX; St. Petersburg, FL; and St. John, New Brunswick, Canada. Of the three, the center in St. John is open 24x7. Collectively, these centers receive a daily average of 35,000 calls and 1,000 on-line forms with questions from customers.

The welcome centers have solid credentials. For the past three years, they have kept up their certification from the Service and Support Professionals Association.

Russell Reynolds, who is at the helm of the company's North American welcome centers, recalls that two overlapping developments within Xerox led him to benchmark: access to on-line support from Xerox.com, and a growing number of products that lent themselves to on-line support.

Reynolds sought perspectives from other firms on first-call resolution, customer satisfaction and employee satisfaction.

"I was looking for validation and course correction," he says.

For example, based on internal surveys, Reynolds knew that Xerox's customers were generally dissatisfied if they stayed on hold for more than two minutes. But aiming to raise the percentage of calls reps answered in 30 seconds wasn't going to be enough to keep customers happy.

Reynolds felt that call centers generally overemphasized measures of efficiency in evaluating their performance.

"The industry was too heavily focused on productivity measures," he says.

Reynolds found a lot of good ideas by attending consortia, led by the Houston, TX-based benchmarking firm American Productivity and Quality Center (APQC), on these very topics.

Reynolds learned of other participants' practices through APQC, which did not identify the companies where these practices came from. For instance, when customers call Xerox and say they haven't heard back from field service engineers about scheduling on-site repairs, the company routes callers directly to engineers. This process of escalating calls didn't originate with Xerox.

"It was an element of someone else's good practice," says Reynolds.

He also discovered that he wouldn't find easy answers to the questions he had first posed. That was particularly true with first-call resolution because companies define the term so differently. At Xerox, for example, first-call resolution strictly refers to calls about technical support.

Instead, Reynolds was able to recognize new methods of supporting customers, like allowing them to go to its Web site to complete support requests, look up information about products and download drivers to printers and copiers.

"Part of our knowledge management strategy is to move knowledge closer to customers," he explains.

The education proved rewarding. In 2000, the APQC named Xerox one of five global organizations that maintain model knowledge management practices; the others were Chevron, Hewlett Packard, Siemens and the World Bank.

Reynolds continues to seek out learning opportunities. Besides visiting call centers associated with APQC, he's in the early stages of researching ways to bolster collaboration among the far-flung welcome centers with the help of the Center for the Study of Work Teams at the University of North Texas.

Offering customers more options from Xerox's Web site is still a top priority, especially as Xerox's products make on-line support more viable.

"Today's technology is much more amenable to customer intervention and remote solutions," says Reynolds.

Support Center Certification

More support centers and outsourcers recognize that besides using benchmarking to compare their operations against those that are similar, certification enables them to hold management to specific standards.

Certification also gives prospective clients "proof" of the quality they demand. That applies especially to companies that are selecting outsourcers.

To receive certification, call centers receive audits from third parties that apply open or proprietary standards. The certifying organizations usually maintain committees that set the standards and update them annually. Some certifiers, such as the Help Desk Institute (HDI; Colorado Springs, CO), offer their members on-line self-evaluations before audits to show how their support centers compare to standards.

To become certified, there are several options depending on the standards and the program. Applicants must document and self-assess their processes, and an auditor accredited by an outside firm checks to see if the center meets the standards by examining and grading the center's assessments and operations.

If the center does not comply with standards, the certifying body's auditors point out where. If the center is far out of compliance with the standards, the auditors reject the center's application for certification. But if the difference between a center's practices and the standards are slight, the certifying body generally awards certification, provided the center makes certain corrections. Several certification programs offer or recommend consultants who advise how to fix those problems.

A certification is only as good as the firms certified to it. To ensure that your support center complies with the standards, the top quality certifying bodies periodically review and re-audit centers. They may also occasionally follow up on complaints by customers and employees who allege that you are violating the standards. In the most extreme cases, the certifying body pulls the certification.

But certification isn't cheap. For example, the SSPA's Support Center Practices (SCP) certification, which Xerox's welcome centers have maintained, costs $25,000. The HDI's audit costs about $14,000, subject to adjustment depending on the complexity of the operation, such as whether it has multiple locations. Even the HDI's self-evaluations, which the organization offers only to members, cost $500.

Therefore, you have to carefully examine whether you need certification. If your competition is offering and/or marketing it, and senior management is saying that customers are beginning to demand it, then you should include certification in your budget or put in a budget request for it.

Among certification programs, some are specific to support centers and others are for call centers in general. Here are a few examples:

COPC-2000

The Customer Operations Performance Center (COPC; Austin, TX) administers COPC-2000 certification, which applies to call centers and support centers.

The COPC comprises top call centers and companies that outsource customer service or support. It audits in-house call centers, service bureaus and fulfillment houses to see if they meet the COPC-2000 standards. The standards, based on the Malcolm Baldrige award we mentioned earlier, cover 29 separate items.

To achieve COPC-2000 certification, applicants select certain employees who receive training from COPC professionals to become registered COPC-2000 coordinators. The coordinators work with the company that is conducting the certification audit. A COPC-2000 audit typically involves two or three auditors who spend three to five days on-site. Each center receives separate certification, which takes between nine and 12 months.

Applicants receive one of four grades: certified (complies with all standards); conditionally certified (compliant on 27 of 29 standards, with minor deficiencies on the remaining two); certification candidate (passes 22 items and the applicant agrees to become fully certified on the remaining items within 12 months); and no certification.

COPC conducts a six-month review and annually awards re-certification.

Certified Operation for Resolution Excellence (CORE) 2000

The CORE 2000 certification from STI Knowledge (Atlanta, GA) focuses on support center structure, strategy, methodologies, systems and technology, perception and performance, staffing and training, measurement, reporting and innovation.

The CORE 2000 certification tests to standards derived from industry best practices and methodologies. STI Knowledge validated the practices and standards with its 25-member World Leadership Team, which consists of support industry leaders, practitioners and consultants.

Certified Support Center (CSC)

The Certified Support Center certification from the Help Desk Institute looks at eight core areas: leadership, policy and strategy, people management, resources, processes, people satisfaction, customer satisfaction and performance results.

Independent auditors award certification to support centers based on 60 standards, each with four performance levels covering the eight core areas. The HDI has a series of questions to evaluate the level of a site's conformity to each of the standards.

An industry standards committee developed CSC by applying existing quality and performance certification methods, such as the European Foundation for Quality Management, The Malcolm Baldrige award and ISO 9000.

To prepare your support center for the audit, the HDI provides its self-evaluation survey on its Web site, www.helpdeskinst.com. After an audit, HDI generates a report that shows how your organization compares to its standards. HDI also offers *Compare* an industry benchmarking tool, from its Web site.

International Standard Organization (ISO) 9001

The ISO 9001 is a part of the ISO-9000 series of quality standards. Among the areas the ISO 9001 standard covers are quality system establishment, documentation, management, infrastructure, monitoring and measurement.

The ISO organization, formed in 1947, is responsible for a wide range of international standards, including ISO 9001, which it amended in 2000. Companies on the previous version, ISO 9001:1994, have to re-certify to the current ISO 9001:2000 standard.

To earn certification from ISO, you have to demonstrate to auditors from an accredited registrar that you have defined and implemented an effective management system. Several accreditation organizations in the US serve as registrars, such as the Registrar Accreditation Board (Milwaukee, WI).

Support Center Practices (SCP)

The SCP, sponsored by the SSPA evaluates support centers on 11 major criteria, which include customer relationship management, customer feedback, corporate commitment and strategic direction.

The SCP also covers areas such as recruiting, screening, career path opportunities, job descriptions, employee feedback, stress management and training. It looks at your performance metrics, including how efficiently your center responds to support requests. It examines how your center directs calls and on-line communication to support reps, and evaluates the effectiveness of automated options like interactive voice response systems and public knowledge bases on your Web site.

If you support tangible items or technical services, the SCP examines quality management, such as how your support center works with your engineering department to identify and fix major problems with your product or service. The SCP also assesses how your support team collaborates with your sales and field service departments to resolve customers' problems and respond to sales leads.

Companies first self-certify each call center and self-assess against the SCP standards. Service Strategies Corporation (SSC; San Diego, CA) administers the SCP certification program, which it created with 35 companies that represent the SSPA's core membership.

During self-assessments, applicants follow detailed steps that outline each area they should document, and they measure their results against industry averages and benchmarks. Such assessments take between 30 and 90 days.

Next, SCP audits your support center on-site. Auditors provide you with feedback from each of the 11 program criteria. SCP provides an audit report that includes final scores and feedback on all elements that didn't comply. The report includes reasons why the center didn't comply and indicates the improvements the centers can make to achieve compliance.

In addition, when a company passes an audit, it receives a benchmark report that provides a detailed comparison of its center's individual scores against all other certified centers.

Support centers either pass or fail audits. Those that do not pass have up to 60 days to correct any deficiencies.

Sorting Out Certification

Whether your support center benefits from operational certification depends on the certifying organization and on your company.

"Certification can be very positive if it provides an appropriate framework for improving processes and services, and enables the call center to identify performance gaps and make significant improvements," explains Brad Cleveland, president of the Incoming Calls Management Institute (ICMI; Annapolis, MD). "There may also be marketing value to having a program's stamp of approval."

Adds Cleveland: "On the flip side, if your company does not have clear objectives, does not commit the required resources, or if the certification program provides an insufficient context for improvements, certification can result in a waste of time and money."

There is competition among providers of certification for support centers, as well as debate about whether support centers even need their own certifications.

SSC president John Hamilton says that specialized certifications like the SCP delve much more deeply into support requests and resolution than general call center certifications. For example, the SCP looks at whether the same data on a company's public knowledge base on its Web site is readily available to support reps.

"Unfortunately, this is an all too common problem when companies don't synchronize the link between the two data sources, and consequently customers may not get the most up-to-date and accurate information that could provide the answers they're looking for," says Hamilton. "The auditors also examine whether companies measure the efficiency of their electronic service delivery programs."

The SCP program looks at first-call resolution rates and seeks and mandates that companies get feedback from customers through surveys.

"Our certification complements the ISO 9000 series," says Hamilton. "The ISO program enables companies to meet their customers' broad enterprise-wide quality requirements while ours meets the specific quality and performance needs of support desks. We also require extensive documentation for our standard as does the ISO."

ICMI's Cleveland says that although technologies in support centers may vary, the best centers apply principles proven in other call center environments. These include establishing service level objectives, forecasting reps' workload accurately, scheduling around workload requirements, routing calls efficiently and adapting to real-time requirements.

"Should [support] desk certifications be much different than certifications for other customer contact environments?" asks Cleveland. "Beyond unique systems and user requirements, absolutely not."

Given the range of opinions, how do you sort out certification programs?

Kathryn Jackson, an associate with Response Design (Ocean City, NJ), advises that you look for a neutral third-party certification and a certification process that includes validation of data rather than a review of data you report yourself. She also recommends that you look into the credentials of the people who developed the criteria for certification.

Jackson believes certifications are not prescriptive. They document the strengths and weaknesses; they don't always present solutions. Yet not coincidentally, some certification organizations have consulting services and technology divisions that can identify and correct problems with your audit for the right price.

One consultant, Chad Burbage, president of BC-Group International (Dallas, TX), questions the validity of obtaining certifications from organizations that offer consulting services.

"A proper certifying body should inspect your call center's process, and see if you comply," Burbage points out. "But if and where you don't, it should be up to you to fix the problems. It is your opportunity, after all. If you need assistance to achieve the fix, the onus is on you to find the right consultant, trainer or vendor, and you should not be pressured by the certifying organization to use their services."

When selecting a certification organization, ask companies that earned them how tough the audits were. Certifying bodies and firms that make it too easy to pass are nothing more than diploma mills.

Carefully scrutinize companies or organizations that seem, according to your gut instinct, to depend a *little too much* on certification, based on their fees and/or the marketing benefits they purport to offer for their membership. You don't want to get ripped off.

As you review the certification organizations and standards, verify that they cover all relevant aspects of your business. Otherwise, you risk applying for a certification that presents you with a nice piece of paper but doesn't show you how to do a better job of supporting customers.

Ten

CHAPTER Ten

KNOWLEDGE DELIVERY
A.K.A KNOWLEDGE
MANAGEMENT

Knowledge Delivery a.k.a Knowledge Management

C H A P T E R Ten

Customer support is about knowledge. Diagnosing problems with knowledge from customers and solving them with knowledge from you the vendor.

Knowledge can be as simple as hearing "the computer won't boot up" and as simple as asking: "Did you plug it in?" Or knowledge can be as complex as determining that "the OS crashed" and replying, in the appropriate and apologetic tone (understanding that the customer has pulled out enough hairs to mimic Kelsey Grammar): "We've consulted with the OS vendor and we're going to write and upload a new application to fix it."

Customer support centers deliver the knowledge: from the customer to you, and from you to the customer. Therefore, 'knowledge delivery' is arguably a better term than "knowledge management" because you can't truly *manage* knowledge. But we also use the term "knowledge management" as it is still the currently commonly accepted terminology.

Knowledge is learning what is, what has worked and how to apply solutions. But knowledge also happens. Knowledge is that 'By Jove I think I got it!' mixture of hard work and/or insight. Knowledge is the rep or the customer realizing that the old solution doesn't apply in this case. Knowledge delivery, therefore, is acting on and applying what has been learned.

Here are the knowledge methods and the issues that impact their delivery.

The Knowledge Triangle

There are three legs to knowledge delivery: knowledge about the customer, knowledge about the product/service they have and knowledge to diagnose and solve the problem. Here's how they work together.

Knowledge about the customer is determining and knowing who they are, what they buy, how much and when and how they use it. This is the essence of what has been jargonized as 'customer relationship management' (CRM).

CRM begins when you begin marketing your product, through any means. You locate a store where it will get sufficient traffic of people interested in buying your goods or service, be it in a mall, an attractive trendy downtown shopping area, or in a catalog, or on a popular Web site. You buy advertising in media that your market pays attention to. You send direct mail, catalogs, e-mail and make outbound calls to customers whose names and contact information you obtained from inbound calls, on-line forms, e-mails, customer service cards, warranty information, and from lists you purchased, rented or borrowed.

In all of these cases you had researched your market. You had developed a market profile of who would likely buy your products. You targeted where you located and to whom you marketed based on geographic and demographic data, i.e. if you sell computers you opened a branch near a college or on the edge of a middle/upper-class neighborhood. Or if you're making and marketing CRM software you bought ads in *Call Center* magazine because it reaches your customers.

When the market responds you then develop individual profiles of customers. You ask questions of them in surveys and track their buying patterns. As you add these profiles you begin to segment them into top, medium and low level customers.

One of the market truisms is that '20% of your customers account for 80% of the sales.' But don't ignore the others or treat them shabbily. As they begin to buy more because their needs and incomes change, if you've in the past given them short shrift, they just might give you the name of a well-known rather warm location to relocate your business to (and even offer assistance in getting you there) rather than doing business with you.

This knowledge enables you to set up your support center to meet the needs of the best and the most promising customers. The usual support operation has a tiered approach for low to mid-level customers (first self-service than a zero out to live reps) and a direct approach for high level customers (direct to live reps). To

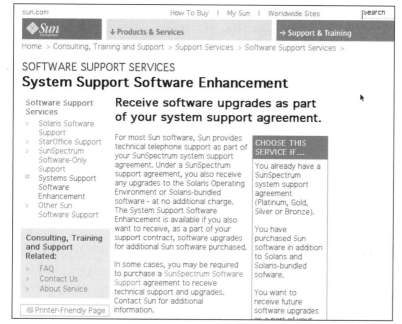

For most Sun software, Sun provides technical telephone support as part of your SunSpectrum system support agreement. Under a SunSpectrum support agreement, you also receive any upgrades to the Solaris Operating Environment or Solaris-bundled software - at no additional charge. The System Support Software Enhancement is available if you also want to receive, as a part of your support contract, software upgrades for additional Sun software purchased.

In some cases, you may be required to purchase a SunSpectrum Software Support agreement to receive technical support and upgrades. Contact Sun for additional information.

Sun Microsystems offers a complete support menu, enabling customers to mix and match packages to meet their needs. Ergo, Sun's support reps know what plans customers have so when they call they can provide service that matches them.

cover support costs and raise revenues (see chapter 3) some support centers are now charging for support, usually by offering different support plans — the more the customers are willing to pay, the quicker the support.

But knowing about the customers helps reps understand the customers, and their needs, which enable them to better serve your customers. If a customer prefers to speak Spanish or French, and the questionnaire or Web form asks this and records the information, then when they call the support center it can route their call to a rep who can speak their language. If a customer is new then the reps can be trained to have more patience to walk them through the knowledge process.

This leads to *knowledge about the customer's product/service*. The more you know about what the customer owns or uses, the better able your reps can fix their problems.

This is why it is critical that your firm obtain serial numbers and warranty information to link the items to the buyers. Even the most mass marketed of products, like consumer software, often have their quirks. You must build equipment or service records to track their experience with your goods. The reps need to know what service level agreements (SLAs) your firm has with customers.

You should also have your reps ask, or find out through tools like remote diagnostics what modifications customers have made to the product or service. As related earlier, co-author Brendan Read found this out the hard way when he installed a software firewall that FUBAR'ed his hard drive, forcing him to hire a computer expert to wipe it clean and *reinstall everything*.

Your reps should know about your products and services. But they should also know about the products and services that go into your products and service, like the OS platform that your application is built on. They need to be taught about where and how these products and services interface, and how to solve problems arising from that interface.

Finally, to solve problems requires reps and companies need to have *knowl-*

Remote Diagnostics

Remote diagnostic tools can be used to perform Internet-based diagnostics, upgrades and evaluations of customers' devices, such as computers, electronics and even household appliances. The information gleaned from a quick peek inside a customer's problematic device allows the support rep to provide a quicker and more cost-effective solution to the problem and/or alter the device's parameters to re-optimize its performance without the need of a costly 'truck roll.'

From a software standpoint the tools must be installed on the customers' devices, then from a hardware standpoint the remote diagnostics tools do their 'magic by accessing the customers' devices through a standard phone line or high-speed connection, which connects the device to the support center's systems through a modem. With the locations physically linked, the support center's systems can act as a terminal in accessing information located within the onboard intelligence of your customer's device.

Remote diagnostics can work in one of two ways: the process is only ini-

edge to diagnose and solve the problem. The essence of this knowledge is realizing what problems arise and devising and either suggesting the answers to the customers or implementing solutions to them, such as by remote fixing, like line repair, shipping bug fixes or if a major hardware or software problem, field service.

This knowledge can be static or dynamic. Static knowledge consists of common problems and solutions. Dynamic knowledge is analysis of unique problems and devising unique solutions, 'on the fly.'

Static knowledge can take the form of IVR or Web FAQs or knowledge bases that can take the form of specialized databases, Web sites or other data repositories, your own or partner vendors', accessed on-line by posing questions or by support reps. Static knowledge can also take the form of remote or self-healing

---

tiated by the customer calling the support center for assistance, or the customer can activate the diagnostic software and the software itself will contact the support center's systems for assistance.

When customers contact the support center with a request for help with a problem, the reps evaluate whether or not remote diagnostics will solve the problems. If it's determined that a remote diagnostics session will be of help, the customer establishes a link up. Once the connection is established, the rep accesses the diagnostic tools residing within the machine to determine the cause of the problem. Once the cause the ascertained corrective action can be taken. The fix may involve the rep and customer working through the solution together, the rep installing new software on the machine or a decision that the product must be repaired by a field service technician or shipped back to the factory for repair.

Remote diagnostics is not only used to analyze problems; it can also alter system parameters to find the optimal settings, the current software version installed and the need for an upgrade. These tools can provide a support center with wealth of information about the company's products and how the customer uses them, but these tools must be used responsibly.

Remote diagnostics is a natural evolution in the ongoing effort to provide cost-effective support to customers who buy high-end electronics, computers and related equipment.

diagnostics in which machinery system finds out what is wrong and suggests solutions or cures itself.

Static knowledge is only as good as the last update. To keep static knowledge valid companies must upload new answers but only after testing and proving them out. Like the old programming adage: 'garbage in, garbage out.'

Static knowledge is only as good as the access points. Your reps and your self-service should have access to the same information. That way everyone is on the same page. Nothing bothers customers more than going through a Web site FAQ, finding the answer and later posing the same question to a rep (when the customer can't access the Web site) who is usually a level-one rep, and either the rep can't answer it or gives the wrong answer. Service Intelligence (Seattle, WA), now Kinesis, found that the occurrence of these flaws all too common in a shocking audit of leading firms' support desks in 1997.

Static knowledge is also only as good as the depth of data, and the links to other data. You should arrange this with your vendor partners to enable your customers and reps to solve problems that arise at the interface of your product and theirs, and vice-versa.

Dynamic knowledge usually takes the form of a typically level-two or higher-end level-one rep who cannot solve the problem by accessing static knowledge. They then think literally out of the box and try different solutions until they find one that works.

It is this dynamic knowledge that forms the base of static knowledge. If reps find that the same devised solution works for many customers with the same problem then this can be put into a knowledge base, on a FAQ page or in a bug fix, or if hardware, into a new component.

All companies start out with dynamic knowledge. The engineer/developer or inventor and their loyal assistants are almost always the best persons to answer questions and to fix problems that occur with their pseudo-Frankenstein. But after awhile many of the questions and answers are the same, so why waste their time and risk their wrath to pick their bat-belfry'ed brains on such mundane issues? Hence the creation of static knowledge.

But as with static knowledge, it is often vital for the reps practicing dynamic knowledge to know how to fix and devise solutions for other vendors' products where they interface with their own. Like a car dealer's mechanic who must know the characteristics of different makers' batteries, tires or parts.

Both static and dynamic knowledge saves the products' vendors the hassle of handling the support issues if your customer support solves the problem, negating the need to have the problem escalated to them or ship the product back to them. And vice-versa. We all must work together in this business.

If your company has outsourced part of its support you must also link your static and dynamic knowledge to the outsourcers. Outsourcers should have seamless links to both your static knowledge and to your upper level support center for dynamic knowledge.

How Knowledge Delivery/Management Evolved

Knowledge delivery/knowledge management has evolved considerably since the early 1990's and since 1996 when former *Call Center* magazine editor Mary Lenz wrote her groundbreaking book, <u>The Complete Help Desk Guide.</u>

To understand why knowledge management is broadening its impact beyond the realm of support, it's instructive to look briefly at the evolution of CRM, which emerged under similar circumstances as knowledge management. (We consider the practice of CRM in greater detail in chapter 11.)

First, let's define our terms:

During the early and mid-1990s, support reps used a category of products called problem management software to keep track of customers who called and the problems they were having.

Near the end of the 1990s, a small group of software vendors began to offer products that incorporated information about customers that related to sales and service, in addition to documenting support requests.

These products were the basis for software that is now at the center of many of today's CRM projects. More importantly, the practice of using information about customers as a tool to improve service matured from a parochial concern of support centers to become a top priority in large corporations.

Knowledge delivery a.k.a. knowledge management evolved under similar conditions as CRM. The same support reps who relied on problem management software at the start of the 1990s were soon also using a category of products called problem resolution software. Once a rep learned that a customer had a problem, he or she would use this software, either during or after the customer support call, to try to locate a solution to the problem.

At that time, the Internet had not yet caught on widely among consumers, nor had a by-product of the Internet, the search engine. Problem resolution, like problem management, was often under the domain of the support department rather than a company-wide effort. The job of selecting problem resolution software fell to support managers, who in the process often had to learn about various methods of cataloging information.

One way, for instance, was to refer to previous examples of solutions that worked in similar situations in the past. As long as a support center placed answers to similar questions or on similar topics within the same groups, and told the reps what these groups were, this historical or case-based approach was usually sufficient. Inference, a developer of knowledge management that Sunnyvale, CA-based eGain later acquired, was among the leading proponents of what it termed 'case-based reasoning.'

The only drawback with the case-based method was that if the support center did not organize answers so that reps could easily find them, or if the support center didn't keep the answers current, the reps were better off transferring callers to colleagues, or placing callers on hold and asking colleagues what the answers were.

The early 1990s was a time when handwriting recognition and speech recognition software was beginning to become available to businesses and wealthy consumers. One of the underlying concepts in the development of this software was that over time, and with enough representative examples, computers could learn to recognize how people wrote, what people said and how they classified information.

This idea still applies today.

Businesses that currently use speech recognition to automate certain customers' calls typically conduct pilots in which the callers essentially "train" the software. These pilots tend to take place all over the US at the same time because pronunciation varies from region to region. Companies also can rely on market research firms to find male and female speakers of different ages to train speech recognition systems. The goal of this training is to ensure that these systems recognize as many callers as possible.

The same concepts applied to the first knowledge management features offered by problem resolution software vendors. By the mid-1990s, support managers had the option of buying problem resolution software that, after a certain period of time, would be able to take a reasonably good guess as to how the support center grouped answers to callers' questions.

One of the staunchest advocates of this approach, The Molloy Group, incorporates its self-learning tools within software from ServiceWare, a leading developer of knowledge management software.

Although groundbreaking, this probabilistic approach did not initially catch on. Unlike speech recognition, the process of establishing categories of answers wasn't as clear-cut, from a user's point of view, as saying words or writing letters. Support managers preferred to organize answers in the knowledge base themselves rather than wait until a computer figured out how to do it for them.

Toward the end of the decade, problem management software vendors began to combine the historical and probabilistic methods, which worked better in tandem than they did alone. Some vendors also began to sell pre-packaged answers to common questions about popular software from leading firms like Microsoft and Novell. These answers, referred to as 'canned knowledge bases,' were useful adjuncts to the knowledge bases support centers created when they aggregated answers to frequent questions from their own customers.

Even before the Internet became widespread in the late 1990s, knowledge bases were valuable tools for support centers. They were certainly not substitutes for training, but they helped fill in gaps, enabled reps to quickly find a consistent set of answers for helping customers, and enabled support centers to retain valuable problem-solving information when experienced reps left.

But the Internet changed everything. The users of knowledge bases were no longer limited to support center reps. Customers could also have access to all or part of a company's knowledge base on-line. Knowledge bases now had to accommodate both customers and support reps.

As was the case at many businesses, and often true in support, a gulf separated on- and off-line operations. Problem resolution software vendors reinforced this condition. When vendors introduced products that reps could access through a corporate intranet or a secure portion of a company's Web site, support center managers usually had to make choices between the Web versions or Windows versions of the same products.

That's no longer the situation. As on-line documents and intranets became more prevalent among businesses, in general, at the end of the 20th century, the inadvertent duplication of Web and non-Web problem resolution gave way to the broader discipline of knowledge management. Tools for finding answers to questions, already available to the public through Internet search engines, became necessary not only for support centers, but also for multiple corporate departments.

Eventually, knowledge management initiatives, like CRM projects, began to take place throughout numerous companies, including some we profile in this chapter.

Knowledge management expanded beyond problem resolution in two other important ways. First, in terms of products, a standard feature among many knowledge management tools is software that allows customers to send e-mail messages to support reps or engage reps in live chat sessions if they have difficulty finding answers on-line.

Also ubiquitous on many Web sites are lists of answers to FAQs. These can consist of questions and answers that on-line visitors access by typing in keywords and phrases or selecting from a list.

The second way knowledge management has evolved is that it has grown in scope from a matter strictly for techies to a set of principles that guide how companies disseminate information about themselves. Like CRM, knowledge management is more than software; it encompasses strategies for sharing information about a company among multiple areas outside call centers or IT.

This is good news. The more advanced the technology, the less you need to know about it. Since knowledge management software is more powerful than its predecessor, problem management software, you don't have to immerse yourself in the details of how the software searches for answers, as you would have had to in the 1990s.

Although knowledge management's domain is growing in the customer service arena, support continues to be its focus for now. Ask call center executives today about knowledge management, and they will most likely direct you to their support centers. Still, the Internet has brought about a fundamental change that gives companies the choice of building knowledge bases that fulfill customers' and colleagues' needs beyond support.

Converting Support Knowledge into Sales and Product Development Knowledge

By diagnosing and solving customers' problems you obtain a wealth of information about customers and how they use their products. Dynamic knowledge enables a company to correct flaws in their product or service. To tap that dynamic knowledge you must enable your engineering and product development staff to access your support center, to listen in on calls. No more institutional silos.

You should also have your support team take an active role in product development and marketing so that any potential issues can be resolved, such as too many customer-assembled parts that result in costly support calls, before the product goes live. When customers experience problems due to such issues, it's the support center that has to literally pick up the pieces.

Both static and dynamic knowledge give you insight into individual customer needs, i.e. back to CRM. If a customer says their dial-up connections are too slow then they might be receptive to a DSL offer. If a customer just moved into a new home they might want to buy a wire-maintenance plan. If a product keeps breaking down then the customer should be offered a new one at a discount rate.

But be careful how you handle this information. Many customer support reps are not good at selling; many identify with the customer and not with the company. You should look into having a sales team follow up with this information or carefully train your reps how to sell, on the basis that they are supporting the customer by offering something that will help them.

But only for those products or services that will do that. No support rep worth their beanie will foist a Version 2.5.2.5 if they think it is a piece of potentially recyclable matter i.e. junk.

Getting Everyone on the Same Page

When delivering knowledge make sure that all channels of your company have access to the same customer information, e.g. like outstanding trouble tickets, warranties and service plans. That includes your repair depot, field support, *sales and retail* and your business partners. CRM software enables this.

That way wherever you interface with the customer and that customer asks about an outstanding problem your employees will know what they're talking about, instead of having the typical glazed I'm done for 'deer in the headlamp' expression. And your relationship with that customer won't get shot to the other side of the Styx.

Daren Nelson, founder of CRM software supplier GWI Software (Vancouver, WA) knows that sensation all too well. Before he started that company he had been a salesperson with a large high-tech company.

Nelson would sell software to a customer, whose contract and details would then be turned over to his company's implementation department. On several occasions he would return to that client only to find there were problems with that software that would prevent him from selling more products to that customer.

"Did I ever look stupid going in there," Nelson recalls. "I'm selling the client products that promise better connectivity and I'm not even understanding the problems they were having with what I'm selling."

Nelson walks the walk. One of his clients, global office furniture maker Herman Miller uses GWI's products internally and externally: linking its business partners.

The company keeps its business partners involved in the resolution of their problems and to accurately track the issues that they face. GWI's c.Support program provides progress notifications by e-mail, which allows Herman Miller's business partners to keep up to date on the progress of their incidents.

Because GWI products use a centralized database, all the users, regardless of their physical location can share data. This is very important for a company such as Herman Miller, Inc., which has multiple manufacturing and sales facilities located throughout the world. In addition, GWI's c.Support program has an integrated knowledge base where the solutions to issues can be archived. Technicians have quick and easy access to these solutions, which translates to quick and easy support for Herman Miller's business partners.

Herman Miller's business partners call the firm's support center when they have a software, hardware, or network issue. The caller is asked to provide a network ID, which brings up the caller's profile in the database. GWI c.Support software shows the technician whether or not there are any unresolved problems already on record for that caller, and allows the technician to create a trouble ticket for any new issues.

The reporting capabilities of c.Support allows support center management to track vital information about their staff, such as the ability of the support staff to solve problems on the first call and the number of calls taken per day.

If the rep is unable to resolve the issue on the first call, the ticket can be routed and sent to Herman Miller's second-tier support. "the escalation features in c.Support enable our staff and business partners to always be aware of the status of an issue," says support center manager Thomas Kennedy. "Our business partners appreciate knowing which technician has been assigned to resolve their problem, and when their issues will be fixed."

Today, the support center has a staff of 23 persons on the 'front lines' with an additional 150 second- and third-level people throughout the company trained in handling support issues through c.Support. The support center provides assistance 24 hours a day, seven days a week to over 5,000 business partners engaged in manufacturing, customer service, sales, and product development.

"Our business partners are thrilled with the new level of technical support they are receiving," adds Kennedy. "The c.Support software was instrumental in helping the support center achieve our goal of resolving 80% of their issues on the first call."

What Do Customers Think?

Static knowledge, in FAQs and knowledge bases, solves most problems. Only a minority of problems requires dynamic knowledge to solve them.

But customer feedback is vital to maintaining a helpful knowledge base. After customers access the FAQs or knowledge bases on Web sites, some companies display on-line forms where customers indicate if the answers are helpful and to what extent. (That's in place, for example, at Linksys, the networking products company we describe in this book's introduction.)

The documents that customers deem most helpful appear at the top of a list of suggested answers to a specific question. Those that customers deem least helpful appear at the bottom of the list or not at all.

Other products, such as those from knowledge management software company RightNow Technologies, generate surveys by e-mail that ask customers about the overall usefulness of a company's on-line knowledge base. One thing to remember about these surveys: after you send one, make sure you have customers' consent before you send more.

Knowledge bases that are publicly available to customers have to appeal to a wide audience. The creation of a knowledge base is an editorial endeavor, and, like anything in print, it fails in its mission if few people want to read it. Yet that is the problem with many applications of knowledge management today.

David Daniels, an analyst with Jupiter Media Metrix, a research firm headquartered in New York City, says that companies that are implementing knowledge management software "have a lot of data but haven't used it to best serve the customer." He also finds that "to date, consumers haven't been embracing nor have they been satisfied with" on-line knowledge bases.

In September 2000, Jupiter and The NPD Group, which conducted consumer surveys for Jupiter at the time, polled consumers on-line who had purchased products from Web sites during the previous year. According to an October 2000 report from Jupiter that analyzed the results of this survey, only 29% of the consumers used on-line knowledge bases, and, of this group, 59% said they were satisfied with them

Jupiter's study also found that the percentage of consumers who preferred to reach live reps by phone or through text chat varied depending on what they bought. For example, 55% of the respondents indicated that they would communicate with reps if they were to purchase airline tickets, and 44% preferred to deal with reps if they had questions about loans or banking. The percentage of respondents who preferred live service dropped to 32% concerning hotels and lodging, and the percentage was in the low single digits for purchases of books or music.

Besides determining the types of products or services a knowledge base should present information about, companies also have to consider the overall benefit of a knowledge base.

Daniels does recommend that companies factor in metrics such as how much they would have had to pay reps to respond to questions from customers if they didn't provide on-line answers. But in determining return on investment, the number of visits to Web pages and cost savings are only some of the considerations. One of the most important aspects of a knowledge management system is a mechanism for providing feedback.

"You've got to look at how satisfied the customer was with that answer," says Daniels.

Companies often measure the usefulness of their public knowledge bases on the number of times visitors to a Web site select particular items from the knowledge base or on how customers rate the answers. Some Web sites display questions that prompt customers to answer "yes" or "no" to whether the knowledge bases helped them, or ask them to rate items in the knowledge base on a scale.

The best way to illustrate the practice of knowledge delivery/knowledge management is through examples. Below we describe what led companies to implement knowledge bases to enhance the live support they provide customers.

We conclude this chapter with recommendations for using knowledge bases effectively and considerations for maintaining a knowledge base worldwide.

Case Study: Cutler-Hammer

Based in Moon Township, PA, Cutler-Hammer makes electrical equipment and is a group within worldwide industrial manufacturer Eaton. Although Cutler-Hammer's primary customers are electrical distributors, the company does offer products, including surge protectors and surge suppressors for residential and commercial customers. The company also provides electrical engineering services.

Cutler-Hammer employs several hundred reps at six customer service and support centers throughout the US and Canada. These sites include a tech support center located at the company's headquarters, as well as a center in

Hamilton, Ontario, Canada, where reps answer calls from Canadian customers, and a center in Houston, TX, where reps receive orders for large projects. All told, these centers receive tens of thousands of calls each month.

Reps typically answer calls between 7 am and 7 pm on weekdays, and the bulk of the calls are to fulfill orders for equipment. Some of the orders, especially those that come into the Houston center from construction sites, can be for between 30 and 40 items at a time, which is why the same reps who receive calls also handle orders by fax and e-mail.

As recently as 1999, first-level reps who didn't know answers to callers' questions would often speak to engineers and then call customers back. The engineers had a lot of information to share and included two people who each had more than 30 years of experience, but the members of the front-line staff weren't necessarily able to answer technical questions by themselves.

"Because we were beginning to see a lot of churn in our customer support organization, we were always in a training mode," says Darrell Johnson, a project manager in Cutler-Hammer's global customer support services organization and a member of its leadership team. "We had no mechanism to collect all of the knowledge that [the engineers] have in their heads."

Johnson, who had a dozen years of experience in customer support, took on the assignment of building up the company's internal on-line knowledge base. He started with Cutler-Hammer as a mainframe programmer, and had been a high school English teacher before then, so he appreciated the challenge of providing information that was well-written, yet technically detailed and accurate.

In January 2000, using Oakmont, PA-based ServiceWare's eService Suite, Cutler-Hammer began building an internal knowledge base for reps at the main customer support center at its headquarters in Moon Township, PA. The company initially relied on training from ServiceWare when it first began using eService Suite, but it now provides its own training on the software. Johnson says that reps usually need about an hour and 15 minutes to learn to search through the knowledge base and to add items using an on-line form based on Microsoft Outlook.

The first items in the knowledge base were based on the most frequent requests from callers. They also included information that, as Johnson describes it, often existed as "marginalia in manuals" among application engineers but was not always readily available to reps.

"One factor that has slowed us down is having to pay such detailed attention to the format and structure of knowledge," he says. "It literally took us a month to put together guidelines and standards for publication."

Since February 2000, Cutler-Hammer has sought to increase the size of the knowledge base. Forty-seven experienced front-line reps contribute to the knowledge base, and ten engineers who have expertise with different types of products edit these contributions, often providing additional information and suggesting links among other entries.

Cutler-Hammer also employs a publisher who sets guidelines for entries, and a systems administrator who provides troubleshooting and support for the knowledge base.

Figuring out how to encourage support reps to contribute to the knowledge base wasn't always easy. In the middle of April 2000, the company began requiring all support reps to contribute ten items per week to the knowledge base on their own time. "We did not give them protected time to do this activity," he says. "The idea of setting performance expectations and then possibly applying the big stick — that's antithetical to building a knowledge-sharing culture."

By the end of 2000, the company relaxed its rule about the number of items reps should provide each week, applying the requirement only when supervisors explicitly provide reps with time during the day to work on the knowledge base. "The quality is so far superior to what we got when supervisors were shaking their big sticks," points out Johnson.

He also tried to sell the company on an incentive program that would offer quarterly payments to reps who contributed to the knowledge base. Cutler-Hammer did not go with this plan. Instead, since spring 2000, Johnson has rewarded reps who provide large numbers of entries or particularly useful information with award certificates during weekly staff meetings. The company prints out these certificates in color and displays them in glass-enclosed wooden frames. Besides public recognition, reps earn the equivalent of between $25 and $1,000 in baseball tickets or vouchers for restaurants, among other items. As many as six support reps receive recognition during the course of a month.

By the end of July 2000, Cutler-Hammer had amassed 2,000 knowledge base entries. To celebrate this milestone, Johnson invited 50 employees who had provided or edited knowledge base items to the company's cafeteria for a "Golden Nuggets" breakfast. He served the reps wearing a white chef's hat and an apron.

Knowledge management has become a high priority within Cutler-Hammer. The company held its first-ever support conference in June 2001, which support managers throughout the company attended. During that same summer, the company began to make ten years' worth of technical information available on-line to its research and development group.

From his experience, Johnson, who describes himself as "a committed advocate for knowledge management [who] reads extensively on the subject," has come to recognize that a company's approach to knowledge management is a barometer of its culture.

"To be successful, I firmly believe that it is very important that you get all levels of people involved," he says. "Information only has value to the extent that you broadcast it. The payoff is in the sharing of knowledge, not the hoarding of knowledge."

Case Study: dynamicsoft

Matt McCabe, dynamicsoft's director of customer service, believes in transcending the dichotomy between live service from reps and self-service, where the customers themselves look up answers to questions.

dynamicsoft, an Internet protocol (IP) telephony software company, offers highly specialized products that allow companies to send calls and on-line communication over the same networks.

"We want to lead customers through knowledge management," explains McCabe. "The goal is that they find their answers, and if they don't, they find a team or person who can answer them."

Founded in February 1998 and headquartered in East Hanover, NJ, dynamicsoft provides software to companies that offer voice-over-IP services such as Newark, NJ-based Net2Phone. (The software relies on the session initiation protocol, which, like H.323, is one of a growing number of protocols or recommendations for voice-over-IP and other forms of on-line communication.)

Instead of a team of dedicated support reps, dynamicsoft refers calls for support to developers, who comprise the majority of the company's employees.

McCabe says that typical calls refer to problems with software, lack of documentation or lack of training on how to use the software. The company's ongoing goal is to drive the majority of support requests through an on-line knowledge base, which is why it began using eServer software from Seattle-based Primus to put together its on-line knowledge base in 2000.

Before dynamicsoft began using knowledge management software, McCabe recalls that "customers were calling development directors and [with] no means to capture knowledge so that it could be re-used."

dynamicsoft's customers now have the option of looking up information themselves, and if they can't find answers, they can click on an icon to generate a voice-over-IP call to the company. Support reps receive the calls first, and if they aren't able to assist customers, they use the knowledge management software to escalate and forward customers' requests to developers by e-mail.

The company rotates the task of answering the e-mail among the developers on a weekly basis. After five weeks, all developers have an opportunity to respond to customers' questions.

Since setting up an on-line knowledge base, McCabe says that as of the end of 2000, 12% of customers' inquiries referred to questions dynamicsoft has previously answered. The company updates its on-line knowledge base whenever developers observe that they frequently receive the same questions from customers.

Although the knowledge base currently contains a wealth of material about SIP, including articles from *Communications Convergence*, a monthly magazine that's also a sister publication of *Call Center* magazine, McCabe has found that the scope of the information isn't necessarily limited to IP telephony; it includes directions to the nearest McDonald's.

Case Study: Citizens Bank

Unlike Cutler-Hammer and dynamicsoft, whose respective products and services are largely industrial and highly technical, Citizens Bank offers a knowledge base to answer questions that consumers generally ask of banks, such as how to set up checking accounts.

Based in Providence, RI, Citizens Bank is among the largest banks in New England in assets. Before it acquired Mellon's retail, small business and mid-size business banking operations in December 2001, Citizens Bank had more than 330 branches throughout Connecticut, Massachusetts, New Hampshire and Rhode Island, and one primary call center in Warwick, RI. (Since the acquisition, Citizens more than doubled the number of branches and expanded its coverage to Delaware, Pennsylvania, New Jersey and Virginia.)

Citizens Bank's call center in Rhode Island is always open, and customers reach it through one toll-free number. As of early 2001, the equivalent of several hundred full-time reps were answering hundreds of thousands of calls per month. The reps include specialty groups who assist customers with questions about investments and on-line banking. The equivalent of 25 full-time reps answer e-mail from customers.

Matt Roach, a customer care manager at Citizens Bank, observes that when customers have questions about on-line banking, the most frequent questions concern how to configure their Web browsers to pay bills or how to view accounts from their PCs. "A lot of time, they just want to talk to a human being," he says.

To reduce the amount of time customers had to wait to receive assistance, the bank began building an on-line knowledge base for customers in March 2000. To maintain the knowledge base, Citizens uses iCARE software from Kana, which is headquartered in Menlo Park, CA.

The information in Citizens Bank's knowledge base includes the bank's hours and descriptions of different types of accounts. If customers call or e-mail the company with a question that is not already in the knowledge base, the rep who originally receives the question drafts a response and refers it to a supervisor. The supervisor edits the response and forwards it to the knowledge base administrator, who is among the higher-level support reps. The administrator then inserts the response into the knowledge base.

Roach says that the turnaround time for adding items to the knowledge base is usually 48 hours from the moment when the rep receives the question to the moment the answer becomes available on-line. In addition to reps and supervisors, others at the bank, such as staff from the compliance and legal departments, also review new entries to the knowledge base.

While working on this book, the authors visited the bank's on-line knowledge base from time to time. The home page of Citizens Bank's Web site greets us with a button near the upper left corner that reads, "Need Help? Ask Citizens." After we click on this button, we see a list of between five and ten of the most common questions. Among these are questions about how to enroll in on-line banking and how to contact the bank.

Visitors to the knowledge base have the option of typing their own questions, and they can restate their questions if the answer is not sufficient.

When we go to the site, the question we generally ask is "What checking accounts do you offer?" We then see a list of answers in order of potential relevance. The first answer provides information about all types of checking accounts and the second answer gives us information about business checking accounts. When we select answers, whether they concern the list of the most common inquiries or a specific inquiry about checking accounts, a question on the bottom of the Web page asks if the knowledge base has helped us.

We then have the choice of indicating that the answer is helpful, of continuing to look up answers ourselves, or of completing an on-line form, which ultimately reaches a rep by e-mail. We are also able to review our searches through the knowledge base, including the pages we visited, the questions we asked and whether we wanted to refer any questions to reps.

In applying knowledge management to customer service, Morin says that Citizens Bank's biggest challenges were not technical but strategic. These challenges, he explains, include "tapping the institutional intellect of the bank" and creating a consistent format for the knowledge base so that the same questions receive the same responses. Citizens Bank also designs its knowledge base to use the same language that customers use in asking their questions.

The most important priority, in Morin's view, is ensuring that customers still feel that the bank is available to help them.

"We're deliberately not trying to drive volume to a lowest-cost channel," he says.

Taking Charge Of Knowledge Management

Peter Dorfman has seen knowledge management evolve from an issue that only tech support departments care about to a requirement for any organization.

The founder and president of knowledge management consultancy KnowledgeFarm (Lebanon, NJ), and a former executive with The Molloy Group, Dorfman finds that in many organizations, implementations of on-line knowledge bases usually emerge from the ground up rather than from on high.

Although he believes that the existence of an executive position such as a chief learning officer within a company "is a really nice ideal," Dorfman often finds that knowledge management projects tend to begin in customer support departments. And, he adds, they're more likely to "grow organically from within a small section of a company partly because customer support people are apt to be proponents of new applications of technology."

Dorfman says that the key criteria for deciding what belongs in a knowledge base should be if customers and reps are likely to refer to the information frequently, if they consider the information critical and if the information makes a real difference to the people who use the knowledge base most often. That is why he prefers that support reps, whose jobs require them to find answers for customers or colleagues, be regular contributors to their companies' knowledge bases.

"It doesn't make sense to hire people who are knowledge authors," says Dorfman.

Besides recommending that companies rely on employees, rather than outside staff, to provide information for knowledge bases, he also advises that companies set aside time for reps to be off the phones so that they can make these contributions.

This suggestion reflects his observation that the two largest issues companies face in setting up knowledge management systems are 'undocumented procedures' and the realization that, in the early stages of building knowledge bases, it is a burden for people who have another job to do,"

"Very few people get paid just to manage knowledge," Dorfman says, adding that those who are in charge of knowledge management projects should be good salespeople and politicians who can convince others — primarily colleagues who do not report to them — of the value of regularly maintaining and contributing to the knowledge base.

Part of encouraging reps to work on knowledge bases involves demonstrating to them that the company and its customers benefit from the reps' contributions.

One of the prerequisites to using knowledge management, says Dorfman, is the ability to define the right criteria for measuring and benchmarking the usefulness of a knowledge base. He observes that certain types of organizations tend to have similar goals in creating on-line knowledge bases.

Priorities vary among companies. In Dorfman's experience, organizations that offer professional services usually expect knowledge management systems to enable them to deal with a wide range of issues. Customer service operations, however, want these systems to reduce costs, and sales organizations believe these systems should help them increase revenues.

But organizations don't always choose the best metrics for determining the effectiveness of their knowledge bases. Dorfman challenges the perception, which he believes is common among many support operations, that it's worthwhile to calculate first-call resolution rates. He argues that first-call resolution is difficult to measure because companies are not always able to pinpoint when or if they fully solve a problem.

He also suggests that companies not build knowledge bases with the sole intention of reducing talk times. By allowing customers to look up answers to simple questions, knowledge bases encourage customers to call reps with questions that require detailed answers. "When you add knowledge management, call time typically goes up," he says.

Instead, Dorfman recommends that companies measure how often reps have to refer questions to others, which often happens if an answer to a questions doesn't reside in a knowledge base. He also advises that companies use surveys to measure whether customers are satisfied with the clarity and accuracy of the answers they're receiving. In addition, he encourages companies to ensure that the information in their knowledge bases remains current.

"The most underdeveloped part of knowledge management tends to be reporting," he says. "The big issues have to do with how knowledge ages; the system should remind you when it should be updated."

Dorfman finds that in the actual process of maintaining a knowledge base, "project management becomes a high-priority skill." Since the overseer of a knowledge management project relies on others to provide the information, it is essential that those who contribute to a knowledge base know how to communicate their expertise to their colleagues and customers.

Besides empowering reps to share their knowledge so that others can apply it, Dorfman also believes in setting precise goals during the course of a knowledge management project, such as defining a workflow for contributing entries or drafting style sheets for the entries. Knowledge management is an ongoing effort, and each stage makes a difference.

"If you want to generate enthusiasm, show people a lot of small successes," advises Dorfman.

Going Global

If you have international customers, then all aspects of your Web site have to accommodate them, including the portion of the site devoted to support.

It's difficult enough to keep a knowledge base current in one language. When your company has to translate and update its knowledge base among multiple languages, the challenge becomes even greater.

Firms like Lionbridge Technologies, based in Waltham, MA, can translate items in your knowledge base into European languages, such as French, German and Spanish, as well as Asian languages, including Korean and Japanese.

Straight translations are often necessary, but you don't want to set up a separate knowledge base for each language. For example, IBM, a client of Lionbridge, maintains its support site in eight languages and generally has between 200 and 300 changes per day in the English-language version alone. The costs for verbatim translations would be prohibitive without automation.

One way to streamline automation is to design your knowledge base to be independent of the language in which you originally create it. In terms of content, you have to think of your knowledge base not only as a collection of static documents or files but also as a database of terminology and sentences. When you have a structure for porting frequently-used terms and sentences among a variety of languages, you can reduce the time for and cost of translation.

In addition to translation services and software for globalizing on-line content, Lionbridge also offers consulting on how to put together a knowledge base for an international audience.

Among the technical factors Lionbridge considers are the formats in which the information for the knowledge base is available and the number of locations where the information resides. Formats can refer to the types of software that companies use to organize knowledge base items, like databases, content management systems or electronic documents. Formats also refer to the particular products companies rely on, like Oracle databases.

As the case studies from this chapter show, nontechnical factors such as the number of languages are key. Other factors include who views the knowledge base: employees, customers or both. Before you make your knowledge base multilingual, make sure you know how often you receive contributions to the knowledge base, as well as how much time you usually need to evaluate the contributions. This way, you can develop a realistic timeline for when your company publishes them and for when your company selects which items should be public.

When using a knowledge base, or other forms of static knowledge as well as dynamic knowledge to solve problems, your reps must record the customer comments and results in their language. But you do not necessarily have to automatically back-translate them into your company's home language, which for most American firms is English. Only if it is a serious problem that must be solved by engineers back home is this necessary. But it is important that you use the same terminology, in all communications no matter the language so that your reps know what they're talking about, wherever they are.

Eleven

CHAPTER ELEVEN

CUSTOMER RELATIONSHIP MANAGEMENT

Customer Relationship Management

CHAPTER Eleven

No matter how many years your company has been in business, call volumes and staff turnover can still be difficult to predict. But certain things remain constant.

One is the need for support. Whether customers buy PCs or insurance, you can always expect them to ask questions that require answers from experts.

Technological advances have enabled companies to devise more efficient ways to give support reps, and where appropriate, their customers, answers to questions about their products and services using on-line knowledge bases.

Data Versus Knowledge

From a broader organizational perspective, a company's most valuable information isn't about its products; it's about its customers. Companies rarely lack this data, but they often encounter problems because useful data isn't readily accessible. Plenty of companies have customers' records residing among multiple departments, like billing, customer service and customer support. Some departments have inaccurate or incomplete records. Or customers' records in one department contradict those in other departments.

That's bad enough. The implications are worse. Let's say customers stop doing business with a company because support reps don't assist them to their satisfaction. If the company does not consolidate customers' records, then it can't get a quick read on whether its support department helps or hinders the company's effort to retain customers. As a result, the company risks not being able to identify the cause of customers' dissatisfaction in time to remedy it, let alone become aware that dissatisfaction exists. A good example of such a disconnect is the Fleischer/MCI story related in chapter 1.

The reason your company collects data about customers transcends the need to find out where to ship products or how customers pay for support. With some types of data, a company can identify its most valuable customers, such as those who frequently buy from the company, or those who represent a lot of revenue through their own purchases and through referrals to other prospective customers.

Other types of data enable a company to find out what its customers value most. If a company can demonstrate that customers prefer particular products, enjoy a certain level of support or respond to specific types of promotions, then the company gives itself more opportunities to ensure its customers' loyalty.

Your support department cannot gather all this information by itself; it's a project that requires the involvement of the whole company.

You can make the case for the entire company's participation in this project. The concept of maintaining relationships with customers exerts its influence on

CRM Defined

As you begin your support center project, there's a good chance someone will utter the phrase "customer relationship management," or "CRM." In the late 1990s, CRM was a vague concept that promised peace, love, eternal bliss and perennially happy customers. Rarely, it seemed, did anyone ask what CRM means or how it relates to support.

Here's an attempt at a definition: CRM refers to an organization's ability to improve the ways it communicates with customers based on information it continually gathers about them.

Beyond this definition, CRM broadens the general mission of call centers, which do not simply handle calls or answer e-mail but exist to help customers.

entire companies, but it begins in call centers. Most likely, your support operation is best equipped to guide this effort because it is more knowledgeable than any other department about customers' questions and problems.

Delivering Knowledge Versus Communicating With Customers

What at first appears to be a problem with communicating with customers can turn out to be an issue with routing customers to the right reps or disseminating information about your company. Before you're ready to think about CRM, you need to have other systems in place.

You need to know how customers are contacting your company, what they want, what information they receive from your company and how they act on this information. Support reps, as well as self-service systems like IVR systems and searchable on-line knowledge bases capture this information and any outcomes that follow.

For example, if customers want to know how to reconnect to your Internet service, support reps need to ask customers if they are trying to reset their passwords automatically. If customers are, in fact, attempting to reset their passwords, but can't, then reps should be aware of this to find out what else might be causing the problem.

You also need to ensure that the information customers receive from your IVR system or Web site is the same as the information support reps look up within or through your corporate knowledge base. This includes verifying that reps at all levels present correct and consistent answers to each question. When problems or disputes occur that they can't resolve, reps have to be able to judge when to escalate issues to the next level.

Nothing annoys customers more than when one support rep says one thing, another rep says something different and your Web site says something else entirely. Customers also have little patience for a company where support reps contradict or plead ignorance about information on the Web site.

Yet this is a common experience for customers of large and small companies alike, as research firm Service Intelligence (Seattle, WA) has skillfully pointed out in its well-publicized audits. Service Intelligence conducts mystery shopping, where researchers visit companies' Web sites to locate answers to certain questions and then, posing as customers, call these companies' support centers to pose the same questions to the reps.

In a 1997 audit, published in *The Wall Street Journal* and broadcast on radio stations including WCBS-AM in New York, and covered in *Call Center* magazine's October 1997 feature article, "Keeping Your Support Center Afloat," Service Intelligence found that reps either had the wrong answers or said a problem was not solvable in 25% of 90 completed calls. Callers sometimes had to wait ten minutes or more before the reps found the right answers.

Service Intelligence's 1997 audit covered six major software companies: Adobe, Corel, Intuit, Lotus, Maximizer and Microsoft. The shoppers asked questions about commonly used home and business software like Excel, PageMaker, WordPerfect and Quicken.

Little has changed in that regard, according to Peter Gurney, manager director of Seattle, WA-based research firm Kinesis. A former executive with Service Intelligence who worked on the firm's 1997 audit, Gurney says that although Kinesis has not repeated Service Intelligence's study, he observes that inconsistency continues to plague support centers.

Practicing CRM

For CRM to emerge as a viable approach to doing business, you first have to ensure all areas of your company aim to support customers rather than operating as though customers exist to support them.

Gurney says that organizations tend to think in terms of individual transactions rather than their overall dealings with customers. The reasons, he says, are numerous: poor planning, departments working in isolation with one another, lack of input from customers and buggy software.

An indisputable fact, which Gurney's firm Kinesis points out in a white paper and the editors of *Call Center* magazine have repeatedly noted, is the idea at the center of CRM: every time a company and a customer interact, the company learns something about the customer.

As more companies attempt to practice CRM, they risk duplicating others' mistakes, like pigeonholing customers based on past actions, or worse, emphasizing a customer's value to them rather than their value to customers.

A company's ability to put CRM into practice depends not only on how easily the company performs certain tasks, like installing software or accessing different data sources, but on how well its departments join together around the same goals. The company we profile in this next section illustrates the challenges of making a CRM project work.

A Safe Place For Customer Data

Before Edison Security became part of ADT, which purchased the company in 2001, it provided security systems, such as motion sensors, closed-caption TV monitors and fire panels, to homes and businesses in California.

As a separate operation, Edison Security primarily served residential customers, who typically leased systems from the company for between two and five years. When customers' alarms went off, they would automatically generate calls to one of two 24x7 monitoring centers. Depending on the locations of the customers, the calls reached either a 150-seat center at the company's San Dimas, CA, headquarters or a 25-seat center in Fresno, CA.

Edison also maintained a customer care center, where agents were available between 6 am and 10 pm Pacific time seven days a week to assist new customers who wanted to schedule appointments to install security systems. They also helped current customers with questions about bills or requests to change their services. On average, the agents answered a total of 1,500 calls per day.

This center was the starting point for a successful CRM project, which started two years before ADT's acquisition.

In spring 1999, Edison relocated the customer care center to a larger, 300-seat site in Las Vegas, NV. At the time, gathering information about customers was difficult. The company had acquired new customers after purchasing two other providers of security systems the year before. Information about some 200,000 of these customers resided in a system that was not even Year 2000-compliant.

Edison also had 42 branch offices that maintained customers' contracts in multiple systems and that sent paperwork about contracts to different locations.

These circumstances led Edison Security to implement Amdocs' (Chesterfield, MO) Clarify software, which allowed the security company to consolidate information about all of its customers.

"We were looking for one single place for customer data," says Mario Provenzano, Edison Security's former director of systems development.

Edison installed ClearCallCenter, ClearSupport and ClearContracts, which are modules that keep track of customers' calls, support requests and contracts.

Systems integrator Zamba Solutions assisted Edison with deploying these modules at the customer care center and at the two monitoring centers. Zamba also helped Edison with implementing the modules at its branch offices and at a call center in Santa Monica, CA, where between 75 and 100 agents provide internal support to field technicians.

Given the scope of this venture, Edison had to plan ahead. The project started in February 1999, and the company began to deploy the Clarify modules in December 1999, completing the implementation in March 2000.

The need for advance preparation tied in closely with Edison's aim of enabling its entire organization to gather information about customers efficiently. But, to do so, it would first have to introduce new processes.

"Without a doubt, the biggest challenge was the redesign process," says Provenzano.

One new process, keeping track of different categories of calls, had ramifications outside the call centers. Edison Security consolidated 42 branch offices into 28 locations, where employees were able to view contracts and other information about customers using the modules from Clarify.

"[Knowing] the type of questions about bills helped us to better define what bills should look like," he says.

The redesign also made a difference in the way Edison updated customers' records after installations of security systems.

"One of the biggest drivers is the amount of time it takes installers to get paperwork back into offices," explains Provenzano. "It needs to be in the system two days after installation is complete. We [were] able to measure that."

He adds that the company, using software from Brio Technology, was able to safely vouch for the accuracy of reports it presented to executives, including reports on service levels, contracts and sales and installation cycles.

Provenzano recognizes that it is not always easy to make changes to long-standing methods of maintaining information about customers.

"Some folks were not necessarily on board," he recalls. "Each group tends to have its blinders."

But because the different groups came together with the objective of sharing information, the project was a positive learning experience for the entire company.

"This process was allowed to educate everybody," says Provenzano. "We've come to see how powerful information can be if we take the time to capture it."

Alas, this story does not have the happiest ending. The acquisition of Edison Security and a slowing economy forced ADT to scale back its CRM efforts and close some centers at the end of 2001 and at the beginning of 2002. After emerging as a champion of CRM within Edison, Provenzano left the company after it became part of ADT.

In an ideal world, CRM transcends economic conditions, and creates a framework for focusing companies' efforts toward learning more about customers. In reality, when a company experiences significant change, whether it's growth or downsizing, it's likely to perceive CRM as a luxury it can only afford when it's doing well.

Meeting the challenges of CRM involves more than a successful initial implementation; it entails a long-term commitment, which is not always possible if the people in charge at the outset of a CRM project aren't around to sustain it.

The Upside of CRM in a Support Center

CRM sounds great in theory, but it can be risky, costly and time-consuming in practice. However, your company can really gain from implementation of CRM within its support environment and then sharing that information with multiple departments (sales, product development, engineering, and, of course, marketing).

First, your support center regularly communicates with customers who are actually using the company's products and services. The support center knows how customers perceive the company and its offerings — what's good, what's bad. That knowledge, which support centers glean from communicating with customers, can help a company improve products. A company can even discover demand among its customers for products that it could, but doesn't yet, offer.

It can also assist support reps in cross-selling and up-selling products and service, but you should be wary about asking your reps to perform such duties. Selling is a job that requires different attitudes and skills than serving customers, let alone providing support. It's difficult enough for *customer service reps* to sell over the phone, so use cross-selling and up-selling in a support environment sparingly. Most experts point out that asking support reps to sell to customers when they call for support can diminish the credibility of both the company and the support rep's advice.

At the same time, the practice of CRM decentralizes access to data without changing the function of each area of your company. Your support team continues to gather, and it remains responsible for, unique data that other parts of your company don't collect.

Departments don't have to possess every piece of information your company knows about its customers, but they do have to be able to view the information they need. What distinguishes the practice of CRM is that the support team and other departments can share what they know about customers with the rest of the company.

Second, you eliminate processes that waste time for your team, your company and your customers.

The practice of CRM means that reps use one piece of software to view all information that relates to customers. Reps don't have to waste time launching additional software to locate records of customers' purchases or of a caller's previous dealings with the customer service department. Nor do support reps have to transfer callers back and forth to other departments.

For managers outside your support team, CRM means that they can retrieve records of support requests by themselves without asking members of your team to run reports for them.

Third, the practice of CRM increases the integrity of your company's information about its customers. Since your support department is responsible for data that concerns support, yours is the only group that can change this data. Anyone else who has access to this data beyond your team can view it but can't alter it.

Similarly, reps in your department should be able to view but not change information that doesn't specifically concern support. If a member of your team creates a new record related to a customer's support request, the software should populate the record with any data the company already knows about the customer that's relevant to the request. The support rep should be able to select from a list of customer items such as contact or billing information for a particular customer, but should not have permission to re-enter or edit this information.

The fewer departments that are responsible for certain portions of a customer's record, the fewer people there are who can contribute to mistakes and the easier it is to pinpoint where inaccuracies come from. Your company is more likely to gain a better understanding of operations that assist customers, including support, if it places greater limits on who owns data about customers than who has access to data.

The Essence of Successful CRM

Once your company recognizes how CRM is a guiding principle that underlies its communication with customers, the company's next step is to create strategies to attain specific goals in helping customers. Software can help you consolidate data about customers and discern trends about how they behave. But you need more than software to enable departments to share information about customers with one another.

If you have not yet discovered this, you're not alone.

"There's been a lack of company-wide strategy," says Sharon Botwinik, a senior analyst with Cambridge, MA-based Forrester Research. "The companies that really failed spent millions in technology but did not change their processes."

Botwinik was the lead analyst for a May 2001 Forrester report, "Organizing To Get CRM Right," which described companies' perceptions of their CRM projects based on interviews with 25 IT managers and 25 senior executives from large, multinational firms in the US.

The study focused on projects that involved several practices within companies; 86% of participants described efforts that encompassed customer service, sales and marketing.

The interviewees had the option of selecting one or more choices from a list of reasons for their companies' projects. According to the report, the largest percentage of interviewees, 58%, said that the motivation behind their endeavors was to maintain one view of customers. The next largest percentage, 28%, said that their companies wanted to satisfy and retain more customers (see chart).

Yet companies often did not measure the ongoing effectiveness of their projects. And when they did quantify what they accomplished, their metrics did not correspond to their goals.

Of the 33 respondents whose companies were integrating new software or in other stages of implementation, 39% had not selected metrics at all. Twenty-seven percent of these respondents limited themselves to internal measurements, like return on investment or call resolution rates. Only 12% applied metrics that related to customers, such as satisfaction, and used metrics related to internal priorities such as efficiency.

The study also highlighted the role companies' organizations played in CRM projects. For example, 40% of all respondents said that the structures of their organizations only had positive effects on their CRM projects, whereas 34% cited only negative effects and 24% experienced a combination of both.

Respondents picked from a list of positive and negative aspects of their organizations. The highest percentage of interviewees, 46%, said that their organizations led different areas of their businesses to maintain the same focus. The next leading positive organizational effect, which 16% cited, was that their companies set up central offices for their projects.

The principal negative organizational effect, which 38% of the participants mentioned, was "lack of coordination and cooperation," followed by "lack of executive support," a problem that 14% of the respondents acknowledged.

Respondents were less enthusiastic about the impact of factors other than the structures of their organizations. Overall, 60% of all interviewees said that nonorganizational issues only exerted negative influences on their projects, 36% observed only positive influences and 2% perceived mixed outcomes as a result of these factors.

Participants indicated one or more explanations for these outcomes, and the largest percentage, 20%, described clear and consistent goals and rewards as a positive nonorganizational influence. The next leading positive impact was support from senior management, which 12% of respondents said their companies received.

The predominant negative impact, shared among 36% of respondents, was "misaligned" goals or the absence of rewards associated with CRM projects. The next most frequent issue, which 20% of interviewees cited, was that they found it difficult "to engage and sell senior management."

These problems are hardly unusual.

"In the past, CRM efforts were very stovepipe in nature," says Botwinik. "There was no way of linking information."

One way to avoid these pitfalls, she advises, is to maintain full-time teams with members of departments that perform different functions. Half of the participants in the study said their companies created teams that represented several departments, which typically included IT and some combination of sales, service and support.

Botwinik recognizes that companies find it difficult to assign dedicated staff to these projects. "Taking people out of their jobs full-time is a very big issue to surmount," she cautions.

But she also observes that in CRM efforts that organizations deem successful, the presence of dedicated teams assures accountability and continuity from the start to the end of the projects. This is especially true when projects enact substantial changes in existing processes or create new processes, such as generating outbound calls when customers visit a company's Web site.

Teams that bridge different areas of companies reduce the risk of duplicate processes and mistakes because they allow these areas to share information with each other. Botwinik believes that teams benefit even more when they gain additional perspectives from firms such as systems integrators and value-added resellers.

When senior executives lead these teams and establish small but clear goals at each phase of the project, Botwinik says, they do more than demonstrate support from upper levels of the company. A senior manager's guidance helps to span the stovepipes.

Botwinik sums up this approach: "Think globally, act locally and deliver quickly."

Twelve

CHAPTER TWELVE

DAY-TO-DAY SUPPORT PRACTICES

Day-to-Day Support Practices

CHAPTER Twelve

What makes activities of support centers different from the practices of other call centers? It depends how you define support.

Companies are expanding their definitions of support beyond helping customers with high-tech gadgets. Support has come to refer to any type of assistance companies provide to customers after they buy their products or services.

But, you ask, if support is simply post-purchase service, then why even have a support department? Why not consolidate support within customer service?

Not so fast. Support certainly overlaps with service, but there is a distinction between the two.

Customers often call or e-mail companies when they have inquiries about the status of processes like the approval of loan applications, or the dates when packages of clothing they order on-line are supposed to arrive. Callers sometimes get in touch with companies even when they are not sure they're ready to buy anything, such as when they ask about hotel or flight reservations. These are requests for updates, but they are not necessarily questions about the products themselves.

A support call, unlike other types of calls, comes from existing customers with specific questions about products and services they have already paid for.

The goal of a support organization is not to avoid calls, even if an on-line knowledge base is sufficient to answer some questions. Also, many principles of CRM originate in customer support organizations. Support — which often entails service level agreements with certain customers who purchase certain products — is at the foundation of CRM.

At the very least, support reps have to demonstrate good-faith efforts to resolve problems within a set amount of time. At best, calls for support are opportunities for reps to offer additional assistance.

Using Technology

While support centers need technology to operate effectively, technology is only a tool. What matters isn't what these various tools do, but how you use them. Technology alone will not enable support centers to put principles of customer service into action. To deploy technology successfully, you and the reps at your center have to understand customers' needs and expectations.

Chances are that your support center is more knowledgeable about customers than anywhere else in your company. If you have reason to believe that's not true of your center, then you have to identify the departments of your company that are more in tune with your customers, and find out which practices in these departments you can apply.

To learn about best practices in support outside your company, in addition to those we describe in this book, good sources of information also include consultants, benchmarking firms, providers of certification and industry publications. We list some of these in chapter 14.

The hallmarks of best practices are small but meaningful goals that lead to particular results, like reducing the time customers remain on hold and raising customer satisfaction scores.

Key Practices

Routing, reporting, automated support and monitoring all refer to processes. In contrast, customer relationship management (CRM),Knowledge delivery a.k.a. knowledge management and quality assurance are practices.

All three practices are crucial to any support operation. The purpose of mentioning these practices here is to explain briefly why they have more to do with the application of technology than technology itself.

As we underscore throughout this book, CRM is the practice of learning more about customers with every interaction with them, and knowledge management is the practice of enabling reps and customers to learn about your company and its products and services, typically by viewing on-line knowledge bases.

Many of the issues companies encounter with these practices have little to do with technology, as we describe in the following chart.

Practice	Typical	Ideal
Customer relationship management	Company perceives CRM project as strictly involving software; puts IT department in charge; project fails; company blames software	Company trains reps to learn more about customers with every communication with them, with the aim of encouraging repeat business, and where appropriate, cross-selling and up-selling — company also sets small but specific goals for each stage of these efforts
Customer relationship management	Company attempts to install software before identifying best practices for serving customers, or seeking multiple perspectives from different areas within the organization	In addition to sharing information about customers, company coordinates its practice of serving customers under leadership of senior management
Knowledge management	Contributions to company's knowledge base come from reps s during their own time, with no additional incentives	Companies offer reps incentives or, at the very least, dedicated time to document customers' most frequent questions and the most likely answers, within the knowledge base
Knowledge management	Contributions come from knowledge base authors who have specialized technical knowledge, but don't know what customers are looking for	Reps contribute to the knowledge base under the guidance of a dedicated knowledge manager and subject matter experts
Knowledge management	Company repeatedly tells customers the knowledge base is an alternative to asking reps routine questions, yet the company does not keep the knowledge base current	Reps regularly maintain an accurate, current on-line knowledge base to help employees and customers gain easy access to information about the company
Quality assurance	Supervisors evaluate reps based on whether they say stock phrases instead of listening to how reps deal with unique circumstances of each call	Supervisors or dedicated quality assurance staff develop consistent performance criteria, and provide reps with detailed, constructive feedback with each evaluation
Quality assurance	Reps receive little follow-up training after evaluations	Reps receive training shortly after evaluations to develop strengths and shore up weaknesses with certain skills
Quality assurance	Reps receive little or no recognition for good or improved performance	Reps have incentives (promotions, prizes and/or company-wide recognition) to demonstrate excellent or significantly-improved performance

When reps use CRM software, they consult a collection of databases with information about customers. The information is relevant within the context in which a rep is helping a customer, whether that context is sales, service or support.

Similarly, when customers or reps consult your company's on-line knowledge base, the information they view is meaningful only as it relates to their questions and it reflects the amount of information your company permits them to receive.

Databases and knowledge bases are merely tools for organizing information. What really matters isn't what software you choose, but how easily support reps, and where applicable, customers, can access and use the information that resides within your database or knowledge base.

Likewise, the call monitoring system you select can be a helpful tool, but its effectiveness depends on your company's approach to quality assurance, which entails the ongoing improvement of reps' communication with customers. Departments within your company, and companies within your industry, are the best places to begin researching methods of recording reps' calls and on-line correspondence, evaluating how reps serve customers, developing a curriculum for reps and rewarding reps who perform well.

You can also learn about successful CRM and knowledge management efforts by finding out about practices in other departments and in other companies. Through your research, you can decide if your center can duplicate the processes that enable these efforts to succeed.

In this book, we illustrate CRM, knowledge management and quality assurance through case studies of companies that apply these practices successfully. You can find case studies within the chapters that cover each of those topics.

Customer Support:
An Evolving Business Model

As we explained previously, CRM has its roots in support. As recently as the mid-1990s, support departments of companies that provide high-tech products found their own standalone database systems sufficient to track customers' problems. But as databases in general became more interoperable, it didn't make sense for companies to maintain distinct customer databases within separate departments, especially if these departments collectively served the same customers.

Technical progress drives organizational progress. It is no accident that the Service and Support Professionals Association (SSPA), one of the largest providers of certification for agents and for support centers, started as an organization of companies that offer technical assistance. It is also no accident that the ubiquity of the Internet helped create demand for agents who can communicate by phone and in writing. Technology and the practice of CRM is what enable customer support to emerge as a skill, as a profession and as a career.

But technology alone doesn't enable organizations to evolve.

Let's refer back to chapter three, which starts with the description of a 1999 group assignment at Pace University's business school in New York City. This project, which the cousin of one of the writers of this book worked on, was a business plan for the creation of a technical support center. (We are proud to mention that Brendan Read's site selection articles in *Call Center* magazine were among the resources to which the professors at Pace referred the students.)

When the students started the project, some first sought to find an overarching business model, but they soon learned that customer support involves more than assigning a certain number of reps to answer phones.

As a concept, support is simple. As an operation, support involves complexities that are not immediately apparent. One of the issues the students at Pace dealt with was whether every rep should be able to answer all types of questions or if they should become experts on specific products. The students quickly realized that this staffing decision would affect other areas of the call center.

For example, before deciding whom to hire, the students had to define what skills reps should have. Then the students had to identify where prospective reps were likely to come from, and they researched locations, such as military bases, where residents would have technical skills. That was only the beginning.

By developing a complete view of how to set up support call centers, the students at Pace learned about an operation that has become a necessity in an era when people prefer to shop or find out information without having to travel to a particular place to do it.

With a fundamental understanding of how to manage customer support, you, too, are building a foundation. You can translate your knowledge of support into the ability to lead CRM initiatives that have an impact on your entire company. Whether you directly run a support center, or you supervise someone who does, you have a lot to look forward to. Today's customer support operations are the model for tomorrow's businesses.

Case Studies

The best way to illustrate how support is changing is to look at some examples. Two of the three companies we profile below primarily offer technical assistance. Yet they are adopting new strategies to help customers, like incorporating support within a larger customer care operation or enabling reps to work from their homes.

After we look at these support operations, we outline how you can apply their some of their practices in your company.

World Access

World Access, a provider of travel insurance based in Richmond, VA, certainly includes operations one would typically find in an insurance firm. The company offers its own travel insurance directly to customers under the name Access America and indirectly through travel agencies. The company also provides overseas medical and health coverage to employees of its corporate clients and on behalf of insurers.

For organizations that issue credit cards, such as banks, investment firms or membership associations, World Access also offers concierge services. Customers who use these services generally call, for example, to request reservations at four-star restaurants, tickets for shows or time on golf courses. As of summer 2001, World Access employed 19 full-time concierges to answer calls and e-mail 24x7.

Josh Chapman, director of World Access' call center and financial services, says that the reps within the concierges group are often busy during the winter, when bad weather can strand customers or their luggage at airports. These reps, or concierges, also experience spikes during the summer, when callers ask about tickets to outdoor concerts.

Sporting events throughout the year generate a lot of calls as well, especially those between the two New York baseball teams. "We had an astronomical amount of people calling to see the 2000 World Series," says Chapman.

Concierges sometimes help customers who have more unusual needs. Chapman says that they once assisted a caller who was searching for a certain type of sandals by locating a cobbler in Boston who reconstructed the sandals from photos.

The concierges speak at least two languages and collectively speak as many as eight. With multilingual concierges, World Access can seek out resources that are outside North America. Chapman recalls, for example, that for one customer, concierges located a jeweler in Italy who crafted a new set of gold cuff links for less than jewelers in the US would have charged.

Regardless of the complexity of a request, World Access expects concierges to inform callers of their timeframes for providing answers. Chapman says that the maximum timeframe is 48 hours after receiving an initial call, even if a concierge is not able to complete a request by then.

"With a support call, there is an action item," says Chapman, explaining how he distinguishes support from service."

The big difference is when a follow-up item is required." He points out that service calls are often requests for information that concierges can fulfill during their conversations with customers. On average, concierges receive between 3,500 and 4,000 calls, as well as several hundred e-mail messages per month. Chapman estimates that up to 70% of customers ask for answers by e-mail after speaking with or corresponding on-line with concierges.

New hires spend their first three weeks in a classroom, and they devote about half of the third week to observing more experienced colleagues as they answer calls. New hires do not answer calls themselves until the first or second week after they complete their training.

Concierges also learn how to use World Access' in-house case management system. Furthermore, Chapman says that the company encourages its trainees to become resourceful, quick and accurate in looking up information on the Internet for customers.

In addition to concierges, World Access employs three senior associates who assign concierges to help customers with individual requests, or cases. There are also team leaders who create schedules for concierges, monitor their calls and provide regular feedback about their communication with customers.

The team leaders use Secaucus, NJ-based Nice Systems' NiceUniverse call monitoring system to record and play back concierges' conversations with customers. When they listen to recordings, team leaders verify that concierges speak with callers in a professional manner and that they understand what customers are asking for. Team leaders also observe if concierges clearly explain how they plan to assist callers and if they are adept at finding out if callers have any additional needs they can help them with.

For concierges with the best call monitoring scores each month, as well as the highest levels of productivity and regular attendance, World Access displays their pictures and takes them out to lunch. The company presents annual awards for concierges with similar achievements during the course of a year.

World Access also encourages concierges to recognize colleagues who assist them with research they conduct for callers. For their help, Chapman says, concierges can earn "World Access bucks" toward ballgames, concerts, movies or merchandise like polo shirts from the company. Concierges also can use their World Access bucks during an annual summer ritual when managers wash concierges' cars.

The company maintains a homegrown knowledge base that mainly consists of Web sites concierges frequently refer to. World Access focuses on gathering links to Web sites with information about restaurants, hotels and entertainment in different cities.

Concierges update the knowledge base three or four times a week, and the company offers financial rewards to concierges who recommend Web sites with accurate information that is useful to their colleagues.

Chapman observes that since introducing the knowledge base in September 2000, concierges have been able to reduce the time they devote to research by 60% within less than a year. He adds that one reason concierges consult the knowledge base is that they contribute to it.

"They helped create it, and there's that buy-in," he says.

Going forward, World Access plans to expand its concierge services beyond customers of credit card issuers to include customers who receive World Access' travel insurance through their employers. Chapman points out that although World Access receives a lot of calls from customers who do not have access to the Internet when they are speaking with the concierges, the company is considering options like text chat on its Web site so that customers can reach live agents when they do happen to be on-line.

Reynolds and Reynolds

Headquartered in Dayton, OH, Reynolds and Reynolds offers software, network management services and consulting for car dealerships.

Randy Selleck, the company's director of customer support, says that Reynolds and Reynolds receives about 90% of support calls at centers in Dayton, OH, and Toronto and Montreal, Canada. The company collectively refers to these centers as the Technical Assistance Center (TAC).

Given the frequency of turnover among staff at car dealerships, Reynolds and Reynolds provides regular training on its systems in addition to technical support. The company also runs management seminars for dealers and helps them assess the types of systems they need, like software for ordering parts or for reviewing credit applications on-line.

As of summer 2001, the company had 240 front-line support reps or associates among the three call centers who receive between 750 and 800 calls per month. Nearly all members of this group answer calls full-time. Selleck says that about half the calls are troubleshooting requests and half are questions about how to use the company's products.

The centers provide 24x7 service only for network maintenance. For other calls, associates are available between 6 am and 11 pm local time on weekdays and between 8 am and 6 pm on Saturdays.

Selleck says that the company hires enough associates to reflect its estimated six-month rate of attrition.

Reynolds and Reynolds also maintains a team of 12 associates whom the company has trained to handle three different types of calls that increase dramatically around the same time each year. These spikes occur between August and November, when dealerships have to clear out inventory at the end of each model year; between December and February, when dealerships complete their annual accounting; and in April, when storms occasionally cause systems to go down.

Since establishing this group in February 2001, the hardware support group increased its service level from 67% to 82% within fewer than six months.

The TAC has earned recognition for its approach to support. It received certification in 2000 from Service Strategies in San Diego, CA, which assesses and audits support operations based on criteria from the Support Center Practices certification program. Based on customer satisfaction surveys, the Help Desk Institute recognized the TAC with its Team Excellence Award, one of the Institute's top honors, in March 2001.

Besides functioning as an operation in itself, the TAC is also a training ground for product managers, developers and sales staff. Six of the eight dedicated authors of the company's knowledge base are former TAC associates.

On average, each manager supervises 25 associates and listens to their live calls rather than playing back recordings. Managers score calls based on associates' demonstration of knowledge of the company's products and their ability to upsell callers on additional services.

Reynolds and Reynolds also factors in statistics such as associates' average talk times and first-call resolution rates. Selleck explains that the company considers a call resolved when an associate provides an answer before the end of each conversation. If a customer calls back about the same problem, the company no longer considers the call resolved.

All new hires receive between ten and 12 weeks of training through an internal program called Reynolds and Reynolds University. Trainees spend the first two weeks in a classroom, where they learn about the company's products and services. During their training, new associates sit near more senior colleagues to observe how they handle calls.

The trainees also have assignments where they solve problems for hypothetical customers.

For the first 90 days after the training, new hires answer calls next to mentors who listen to their conversations.

To ensure that new and veteran associates have easy access to information about products and methods of solving problems, Reynolds and Reynolds developed a knowledge base using software from eGain in Sunnyvale, CA. Reynolds and Reynolds expects all TAC associates to provide at least three entries to the knowledge base every three months.

Senior associates review these contributions. In July 2001, Reynolds and Reynolds made this knowledge base generally available on its Web site.

Associates at the TAC often receive nontechnical requests, like inquiries about bills. Selleck says that is why, in summer 2001, the company began a plan to introduce a single toll-free number for all callers to replace several dozen toll-free phone numbers. From the one number, Selleck explains, Reynolds and Reynolds intends to direct customers who need technical help to the TAC and route callers with other issues to other groups. Selleck describes this eventual operation as a "customer service offshoot."

Reynolds and Reynolds is also increasing its emphasis on cross-selling and upselling. The company seeks suggestions from associates, who, for example, at the end of 2000 came up with the idea of a service that helps dealers with end-of-year financial statements.

"Like most companies, everyone is focused on driving growth," says Ed Bolka, the company's vice president of customer education and continuous support.

Bolka estimates that about half of the company's revenue comes from services that originate with maintenance contracts. He says that associates within the TAC recognize that support requests present opportunities for them to suggest new products and services. That's one reason associates' evaluations take upselling into account.

"We've sold several million dollars based on that model this year," says Bolka, referring to sales in 2001.

Bell Tech.logix

Based in Indianapolis, IN, Bell Tech.logix's services include technical support by phone and on-line for its clients' employees and customers.

In late March 2001, the company opened a site in Indianapolis, IN, which it refers to as a technology center. This site complements Bell Tech.logix's support center in Richmond, VA. In summer 2001, the two centers together employed 135 full-time analysts (reps). On average, each analyst receives a total of 550 calls and e-mail messages per month.

Bell Tech.logix provides nearly all of its technical services to corporate employees and customers. If analysts at Bell Tech.logix are not able to answer questions from a client's customer, the company directs the customer to other firms that support the client or to the client itself.

"Our objective is to provide good support and service concurrently," says Marc Flood, Bell Tech.logix's director of IT outsourcing. "When we escalate, we become advocates for the customer. Ultimately, we take the call and we're responsible."

Bell Tech.logix employs analysts with a range of experience in technical support, which is why the period of time new hires devote to training varies between three and seven weeks. New analysts spend a minimum of a few days in a classroom, where they learn about the company. They also engage in role-playing exercises to prepare them for the next phase, when they sit next to veteran analysts and listen to them on the phone with customers.

Trainees initially assist the analysts with entering information onscreen, such as updates to trouble tickets. After the new analysts show they can document support requests or look up information in Bell Tech.logix's on-line knowledge base, they speak with customers themselves under the guidance of the more experienced analysts seated next to them.

The company allocates between 80 and 120 hours of training time for each analyst per year. Analysts choose courses based in part on recommendations from their supervisors.

Bell Tech.logix furnishes analysts with courses from SmartForce, a Redwood City, CA-based provider of computer-based training systems. Flood reports that the most popular courses among analysts are those that show them how to assist customers with Windows 2000 software and how to use equipment from Cisco.

Supervisors, who manage an average of ten analysts, generally evaluate a random selection of five or six calls per analyst per month. The supervisors also review transcripts of analysts' e-mail messages.

To record and play back calls, Bell Tech.logix relies on Interaction Recorder, which is software that Interactive Intelligence, located in Indianapolis, IN, offers with its Customer Interaction Center (CIC) communications server.

Bell Tech.logix installed CIC at its centers in Indianapolis and Richmond.

Analysts' evaluations partially reflect their contributions and use of the company's on-line knowledge base, which Bell Tech.logix maintains using eSupport knowledge management software from Seattle-based Primus.

The company expects analysts to create new items for the knowledge base, especially if they mainly provide support for new clients. Bell Tech.logix also expects analysts to look up and update items within the knowledge base that concern existing clients.

Some of Bell Tech.logix's clients incorporate portions of the knowledge base within their intranets and share this information with their partners and customers. All analysts and their supervisors are able to view and comment on their colleagues' suggestions for updating the knowledge base, including a team of analysts and managers who implement the company's knowledge management system.

The company's approach to knowledge management is a reflection of the career choices it offers to analysts. Flood says that Bell Tech.logix prefers to designate analysts as subject matter experts if they are knowledgeable about certain products instead of promoting them to supervisors automatically. Subject matter experts set standards for training analysts and they approve changes to the knowledge base before Bell Tech.logix shares updates with its clients.

"We provide a dual path for subject matter experts," Flood explains. "We don't want to push them in to a management role."

He does acknowledge that the presence of the knowledge base has changed expectations for analysts.

"When knowledge bases didn't exist, we really valued individual knowledge," says Flood. Now, he points out, analysts demonstrate their value in their ability to resolve problems.

Bell Tech.logix shows its appreciation for analysts by creating a relaxed and congenial atmosphere. At the new technology center in Indianapolis, which features a foosball table, analysts can wear shorts during the summer and enjoy pork chop sandwiches from their managers during cookouts.

The company plans to expand further without building more centers. Its next goal is to employ analysts from locations outside the two centers, such as from analysts' homes. "Our professionals could be anywhere across the globe," says Flood, adding that remote support will enable the company to assist more clients.

Applying Best Practices in Your Support Center

The case studies outlined in this chapter are based on companies which represent different industries and have different types of customers; but many of the principles that underlie their day-to-day operations are essential in any support center. Here are some of the most important.

Evaluate Reps' Communication with Customers

A call monitoring system is necessary to ensure reps in any call center give correct answers in a professional manner. In a support center, where customers often pay to receive support over the phone, monitoring is a necessary part of your day-to-day operations.

Monitoring calls involves listening to a random sample of recordings, drafting written evaluations of these calls and sharing evaluations with reps. This work is often time-consuming, which is why it's a good idea to set aside time for supervisors, or hire a dedicated staff, to monitor calls.

Numerous companies offer call monitoring equipment and software, as well as off-site monitoring services. *Call Center* magazine frequently covers these companies and the organizations that rely on their products or services.

No matter what system or monitoring service you use, it's a good idea to establish a simple and consistent scoring system for all reps, no matter what products they support. If there is turnover among supervisors or quality assurance staff, you reduce the learning curve for the people who take over. In addition to call monitoring equipment vendors and providers of monitoring services, some of the training and staffing firms we mention in this book can assist you with creating scoring templates.

Set up a plan to evaluate a certain number of reps' calls every week or month; the sooner reps hear feedback, the sooner they can act on it. Offer incentives, like prizes or public recognition, for reps to perform well or to demonstrate improvement on evaluations.

If you want to evaluate what reps are doing on their computers during calls or if you plan to review reps' on-line correspondence, one option is to use software to capture screens. Many call monitoring vendors provide modules, at an additional cost per seat, that trigger screen recordings at the start of calls or as reps use e-mail or chat software.

Provide Regular Training

Many support centers offer courses through their companies or with the help of outside training firms. As you read in chapter 8, some expect reps and managers to pursue certification through organizations such as the Help Desk Institute, the Service and Support Professionals Association or STI Knowledge.

From a day-to-day perspective, courses don't always require reps to leave their desks to attend classes. When your center is not currently receiving lots of calls or on-line requests for support, you can push on-line courses, replete with simulated conversations and exams, to reps' computers. If you'd rather set aside dedicated time for courses, you can schedule periods of the day when reps don't deal with customers.

Whether you choose on-line courses, live instruction or both, make sure you devise a curriculum that reflects both the types of calls reps are likely to receive and the products reps support. Tie the selection of courses to reps' evaluations. As with call monitoring scores, give reps incentives to do well in their training.

Offer Multiple Career Paths

Bell Tech.logix recognizes that an experienced support rep isn't always the best choice to be a supervisor. That's why it presents two options for advancement: reps can focus on becoming experts in particular areas or on developing skills as managers.

At Reynolds and Reynolds, for example, reps can become knowledge base authors or members of a group that answers a specific set of questions at certain times of the year.

For day-to-day operations, multiple career paths means that customers can reach a group of people who devote themselves to assisting customers with difficult or specialized questions rather than supervising other reps.

When you create a group of subject matter experts, keep in mind that some reps, even with new challenges and responsibilities, will eventually want to pursue positions within your company that are outside the support center. Find out the annual rate of this internal attrition and factor it into your headcount.

Reduce Dependence on Experts

The more you can consolidate reps' collective knowledge into knowledge bases, the less dependent you are on a few experts. As we explain in previous chapters, on-line knowledge bases make each individual rep as knowledgeable as all of his or her colleagues combined. That's especially true the more you encourage and reward reps who contribute regularly to a knowledge base.

For day-to-day operations, on-line knowledge bases give customers an alternative to calling reps about questions that have simple and immediate answers. Knowledge bases also enable reps to locate answers quickly when customers do contact them.

Thirteen

CHAPTER THIRTEEN

DOWNSIZING YOUR
SUPPORT CENTER

Downsizing Your Support Center

C H A P T E R **Thirteen**

Can It Be Saved?

If your company's prognosis is not good there are several ways of responding. The first step is to look at ways to save both your support center and your company money. Shift many of your lower-valued no-brainer support calls to Web self-service but ensure that you put in clear links to your reps, like 'call me' buttons and prompts for live support if the customer is on a page for too long.

Look at adopting cost-savings and productivity-enhancing methods like Web-based recruiting and contracting out to staffing agencies. They will lower recruiting, training, overtime and productivity costs by cutting back on turnover while providing superior customer-retaining performance.

Nothing lives forever, not even customer support centers. So when faced with the inevitable, how do you downsize it effectively, while leaving the people with some dignity — so they won't sue you?

If you go the staffing agency route make sure you avoid potential us-and-them hassles. Be certain that agency-staff and in-house agents have a clear understanding of who their employer is. The agency must take responsibility for all personnel issues related to their staff.

Examine your turnover. As discussed throughout this book, there are ways to keep your reps longer by improving their facilities and by offering career advancement, training, certifications, teleworking, incentives and rewards.

Also, examine where you can relocate your customer support center to cut costs. As chapter 6 reveals, labor cost differences between communities and countries translate into large dollar savings sufficient to outweigh relocation costs.

Equally importantly, look at ways of charging for support (see chapter 3) if you don't do so already, or find more effective and profitable methods of obtaining fees, like switching from 900 numbers to 'Plutonium', 'Uranium' and 'Cesium' pre-paid plans. Switching your support center to a revenue-generating 'profit' center from a profit-draining 'cost' center should bring a smile on your CFO's face, and prompt him or her to put away the red pen that was about to be applied to your budget.

You should also make the business case for your support center to senior management. Show that it is a revenue-making and retaining center. Take a look at where your center could add revenue and cut costs. This is especially important for customer support centers that cannot charge for support because of valid competitive reasons.

The Tough Decisions

When facing staff cutbacks, do so by attrition where and when possible. This method has less impact on morale, hence productivity, service and customer retention while sending the message that the remaining team members need to work harder to save their jobs.

Make sure you include your staff in cutback discussions. To preserve their jobs they may be willing to work more hours or take on new projects. "By getting their buy-in you may avoid problems like badmouthing your company to people and customers, and theft if people have to be laid off," advises Rosanne D'Ausilio, president of Human Technologies Global (Carmel, NY). "At least they would leave feeling that they tried."

If you decide to let go employees you will need to have a fair redundancy policy in place to withstand legal challenges. The most common method is seniority — last hired, first fired. The problem with that method is that it doesn't measure quality. The remaining workers may be poor performers compared to the ones terminated.

The other fairer but trickier method is by performance. Fair in that you let the worst performers go. Tricky in that you need to have a reasonably objective and consistently applied method of evaluating reps, supervisors and managers.

Sometimes it's better to close a support center completely because it's worth more to another company. For instance, a new call center may be more willing to pick up the lease. This can be a particularly attractive proposition when the center's staff is still fresh and available to use, than if the staff has been spread to the four-corners of the area. Also the facilities itself is more appealing to a purchaser, and if the interested party is a call or support center, they would find the property of more interest if the equipment is still in operation — not ripped out and sold.

This approach may be fairer to employees, allowing them to obtain other employment instead of hanging on vainly in hope. Not even supports centers pay enough to merit such wishful optimism.

With support centers often having higher-and-fatter-than-usual wiring and furnishings you may find firms wishing to locate software and hardware development and machine-heavy back office processing functions interested in that space. If there is plenty of parking, that might appeal to them too. There are specialized site selection firms that have databases of just-closed call centers and support centers.

There are no guarantees. Even the best support centers and their employees may not be saved. If the economy is down nationwide and locally, or customers don't buy your product or service because it is lousy and/or priced too high, driving down demand, there is nothing you can do except close down and lay off your people.

Once you decide to pull the plug on a support center, ensure that your diagnosis is correct. If you draw the sheet over it and it revives, your reputation as an employer has just gone into the Great Beyond. It does not look good to give your reps pink slips and then call them back three months later. You may not get them back because they won't trust your company's management.

Have a Plan

Because closing a support center is inevitable, you should plan for it at the beginning. But first, you must determine exactly what it is that you want the desk or center to do. Then put in efficient, proven and flexible knowledge management and call and contact-handling technologies and decide which functions you want to have handled by self-service, outsourcers, and teleworking. Planning for the right space and property, putting it in the right community and staffing it efficiently and effectively eases the burden of change.

At all stages put in expansion and exit provisions to cope with change. Look at applications hosting for your technologies, where possible. Put in expansion and exit strategies in your leases, like subleasing. Pick attractive buildings and locations to draw in good people and flip them if you need to close down. Devise, with your HR and legal departments fair hiring and layoff procedures.

Fourteen

CHAPTER FOURTEEN

RESOURCES GUIDE

Resources Guide

CHAPTER Fourteen

We've referenced various resources throughout the book. To help you in your quest for the best support center for your company, we provide you, our readers, with a listing of associations, consultants and organizations that are experienced and qualified to help you with setting up and managing various aspects of your customer support.

Although, we also referenced products, outsourcers or staffing vendors where appropriate, we have not included these companies in our resources guide since many of the consultants and firms mentioned below know and can advise in the sourcing and selection of those products and services.

Association of Support Professionals
617-924-3944
www.asponline.com

The Alter Group
800-637-4842/847-676-4300
www.thealtergroup.com

Banks and Dean
416-746-0444 or 888-241-8198/262-240-9400
www.banksanddean.com

The Boyd Company
609-452-0077
www.theboydcompany.com

BC Group International
214-821-7962
www.bcgroupinternational.com

Philip Cohen AB
011-46-910-199-88

Great Brook Consulting
978-779-6312
www.greatbrook.com

Engel Picasso Associates
505-341-0001
www.engelpicasso.com

Hahn Consulting
503-248-0262
www.hahnconsulting.com

Help Desk Institute
800-248-5667
719-268-0174
www.thinkhdi.com

J. Heacock and Associates
303-841-8799

Human Technologies Global
845-228-6165
www.human-technologies.com

Incoming Calls Management Institute
800-672-6177
410-267-0700
www.incoming.com

International Telework Association and Council
202-547-6157
www.telecommute.org

Kinesis
206-285-2900
www.kinesis-cem.com

Kingsland Scott Bauer Associates
888-231-5722
412-252-1500
www.ksba.com

Kowal Associates
617-521-9000
www.kowalassociates.com

Knowledge Farm
908-236-2849
www.knowfarm.com

Loyalty Factor
603-334-3401
www.loyaltyfactor.com

Oetting and Company
212-580-5470
www.oetting.com

PacTac Advisors
212-755-1687
www.pactac.com

Performance Consulting
508-650-0770
www.performance-consulting.com

Bruce Pollock
877-805-5318

The Radclyffe Group
973-276-0522
www.radclyffegroup.com

RWK Enterprises
303-823-6448
www.rwkenterprises.com

Service Strategies Corporation
858-674-4864
www.servicestrategies.com

Service and Support Professionals Association
858-674-5491
www.supportgate.com

Trammell Crow Call Center Site Selection Group
214-979-6193
www.trammellcrow.com

Suggestions for Additional Reading

The goal of this book is to give you the background on what it takes to run a customer support operation.

But that's only the beginning. Here is a selected list of books and magazines that can further your knowledge of customer support, call centers and related areas.

Technical Support

Lenz, Mary. The Complete Help Desk Guide, Flatiron Publishing, 1996.

Wooten, Bob. Building and Managing a World Class IT Help Desk, The McGraw-Hill Companies, 2001.

Call Centers and Customer Service

Bodin, Madeline and Keith Dawson. The Call Center Dictionary, CMP Books, 2002. 3rd Ed.

Call Center magazine.

Dawson, Keith. The Call Center Handbook, CMP Books, 2001. 4th Ed.

Read, Brendan. Designing the Best Call Center for Your Business, CMP Books, 2000.

Waite, Andrew J. A Practical Guide to Call Center Technology, CMP Books, 2001.

Titles on Topics Connected With Customer Support and Call Centers

Balentine, Bruce, and David P. Morgan. How to Build a Speech Recognition Application, EIG Press, 2nd Ed.

Communications Convergence magazine.

Fink, Donald G. Computers and the Human Mind, Anchor Books, Doubleday and Company, 1966

Grigonis, Richard. Computer Telephony Encyclopedia, CMP Books, 2000.

Laino, Jane. The Telephony Book, CMP Books, 1999, 3rd Ed.

Newton, Harry. Newton's Telecom Dictionary, CMP Books, 2002. 18th Ed.

Reynolds, Janice. A Practical Guide to CRM: Building More Profitable Customer Relationships, CMP Books, 2002.

Reynolds, Janice. A Practical Guide to DSL, CMP Books, 2002.

Reynolds, Janice. Logistics and Fulfillment for e-business: A Practical Guide to Mastering Back Office Functions for Online Commerce, CMP Books, 2001.

Reynolds, Janice. The Complete E-Commerce Book: Design, Build, & Maintain a Successful Web-based Business, CMP Books, 2000.

Wolfe, Tom. The Right Stuff, Farrar, Straus and Giroux, 1979. (America's first astronauts proved that you need breathing, logical and cool professionals to solve problems when things really go wrong)

Glossary

The following is a short glossary of support terms mentioned throughout this book. If a definition contains a term within this glossary, that term appears in italics.

If you're looking for definitions of terms that apply to call centers in general, we recommend <u>The Call Center Dictionary</u> by Madeline Bodin and Keith Dawson. To learn more about call centers beyond the definitions, consult <u>Designing the Best Call Center for Your Business</u> by Brendan Read, the co-author of this book, and <u>The Call Center Handbook</u> by Keith Dawson.

Call center — An operation dedicated to assisting customers who call an organization or visit the organization's Web site.

Case — The entire history of a support request from the beginning to end (or least to the present, if the case does not yet have a resolution).

Customer relationship management — The practice where a company seeks to learn more about customers every time it communicates with them. Companies that follow this practice seek to apply what they gather about customers to serve them better during subsequent interactions with them.

FAQs — Stands for "frequently-asked questions." A list of the inquiries a company receives most often from customers. On a Web site, FAQs appears within a list, and a visitor selects the question that most closely matches his or her own.

After the on-line visitor chooses a question, the Web site displays a list of possible answers or a list of links to documents that contains possible answers.

Various software products help support centers generate on-line FAQs. The ranking of the questions usually reflects the number of times customers choose the questions within a given timeframe. Some products allow you to delete questions customers haven't asked, say, within the past month, no matter how many times customers viewed these questions before then.

Other factors prevail when ranking answers to questions. Various products have their own methods of ranking answers, but one criterion they tend to share is that they let customers indicate if the information within a proposed answer is helpful. The greater the number of different on-line visitors who describe an answer as helpful, the higher the answer's ranking.

FAQs are most effective if you enable customers to search for topics before presenting them with lists of questions. Otherwise, customers have to winnow through unwieldy lists before they locate questions that are most similar to theirs.

In the context of *knowledge delivery / knowledge management* FAQs represent but one way to present answers to customers' questions within a *knowledge base*.

Field service — Also known as field support, field service is the practice of physically fixing customer's problems on their premises or on hard equipment that the customer uses, such as telephone lines.

First-call resolution — The goal of answering a customer's question or solving a customer's problem with a product or service the first time the customer gets in touch with a company.

For a discussion of the challenges of applying this metric, read the chapter on benchmarking and certification.

Help desk — Up until the late 1990s, referred to any department that provided technical support to employees within an organization or to customers of an organization. But with the evolution of support centers since the late 1990s, the definition of a help desk has narrowed to refer to corporate departments that guide employees through resolving problems with computers and printers. By contrast, support centers assist customers, and not necessarily only with technical products and services (see *support centers* below).

Knowledge base — A database of answers to customers' questions about a company's products or services. Knowledge bases used to be the sole domains of support reps until the emergence of the Internet enabled customers to look up answers to questions themselves.

You can determine which information should be public, available to particular customers or only available within your organization. For example, if your company provides support on behalf of various businesses, you're best off segment-ing knowledge bases so that they're only accessible to the customers of each of these businesses. If support reps need to consult certain detailed documents that are not likely to make sense to your customers, it's a good idea to permit the reps within your company, and not customers, to view them.

Businesses have different approaches to deciding who contributes entries to knowledge bases, who edits these entries and who can view them. We describe these and other issues in the section in this book on *knowledge delivery /knowledge management.*

Knowledge delivery, a.k.a. knowledge management — The practice of setting up information about an organization, usually its products and services, so that employees and customers who need to retrieve this information can easily do so.

Level — How organizations rank *support reps* based on the specificity of their expertise. Lower-level reps tend to have less experience and training than higher-level reps, but they help customers with a wider range of products than higher-level reps, and they're usually the first individuals within a *support center* whom customers reach.

Repair depot — A facility where mechanics and technicians physically fix products. If a product's problems can't be fixed by a customer, with or without the help of customer support, by field service/support or by a dealer, the product is then taken or shipped to a repair depot.

Repair depots are the 'OR' of support. If a product can't be repaired there its plugs are pulled and the guts donated to engineering, sold to the scrap dealer or recycled.

Self-service — Allowing customers to use an automated system to locate answers to questions about products or services. Companies offer this option in addition to enabling customers to request support from live reps.

Support — The practice of assisting customers after they purchase products or services. Unlike customer service, support does not refer to issues outside the immediate scope of a company's product or service, like the status of shipments.

Until the end of the 20th Century, support primarily meant help with high-tech products. But companies in numerous industries have come to recognize that they provide support whenever customers have questions or problems that concern their products or services.

Support center — A department within an organization whose main function is to provide support. Like the term *call center, support center* is potentially misleading because it can refer to a group of multiple sites, or a group of people scattered through multiple locations. Nevertheless, this book employs the term 'center' to mean a centralized support function, even though the people who perform this function don't necessarily work from a central place.

Support rep — An individual who provides support for customers on behalf of an organization.

Trouble ticket — A record of an employee's or a customer's request for support. Up until the advent of the Internet, and still to a large extent today, support reps created these records after speaking with customers who called to ask for help. But more companies are enabling customers to create their own tickets and check on the status of their requests for support on-line.